Electroweak Baryogenesis (Second Edition)

An introduction

Electroweak Baryogenesis (Second Edition)

An introduction

Graham White
University of Tokyo, Tokyo, Japan

IOP Publishing, Bristol, UK

ISBN 978-0-7503-3571-3 (ebook)
ISBN 978-0-7503-3569-0 (print)
ISBN 978-0-7503-3572-0 (myPrint)
ISBN 978-0-7503-3570-6 (mobi)

DOI 10.1088/978-0-7503-3571-3

Version: 20220401

IOP ebooks

British Library Cataloguing-in-Publication Data: A catalogue record for this book is available from the British Library.

Published by IOP Publishing, wholly owned by The Institute of Physics, London

IOP Publishing, Temple Circus, Temple Way, Bristol, BS1 6HG, UK

US Office: IOP Publishing, Inc., 190 North Independence Mall West, Suite 601, Philadelphia, PA 19106, USA

Contents

Preface

The next generation of colliders, gravitational wave detectors and searches for electric dipole moments will all converge on answering the question of what happened during cosmological electroweak symmetry breaking. A compelling paradigm known as electroweak baryogenesis suggests the possibility the Universe boiled during electroweak symmetry breaking and as a result we were saved from complete annihilation. Specifically, strange interactions with bubbles in the early Universe created a tiny imbalance between particles and anti-particles that explains stars, planets and physicists. This wild environment of physics that is far from equilibrium is quite a challenge to understand. Therefore, this book aims to cover a pedagogical gap by giving a step-by-step explanation of the concepts involved at a graduate level.

Acknowledgements

I would like to thank Michael Ramsey Musolf, Csaba Balazs, Bjorn Garbrecht, Mariek Postma and many other collaborators for how much I learnt about this subject through various projects. I am also grateful for my family for their endless patience in the production process of this book.

Author biography

White Graham

Graham White grew up in north Queensland, Australia. He developed a love of physics after reading some popular science books as a teenager and decided he wanted to become a physicist shortly after. He has a particular passion for early Universe cosmology as this field is foundational to our self-understanding and the gaps in our understanding give the most compelling reasons to search for new physics. Graham White is senior postdoctoral fellow at IPMU in Tokyo where he currently lives in Melbourne with his wife and two sons.

Chapter 1

Introduction

The origin of the baryon asymmetry of the Universe (BAU) is one of the deepest shortcomings of our understanding of particle physics, as it cannot be explained within the Standard Model. For instance, one cannot simply set the baryon asymmetry as an initial condition as this would be washed out by inflation[1]. On the other hand, coinciding estimates of the baryon asymmetry using different techniques are a triumph of modern cosmology. The baryon asymmetry can be estimated by the deuterium abundance and from the cosmic microwave background (CMB), where the relative sizes of Doppler peaks in the temperature anisotropy are sensitive to the BAU. These two methods give the overlapping estimates of the baryon-to-entropy ratio[2] [2–4]

$$Y_B = \frac{n_B - \mathrm{nB}}{s} \approx \frac{n_B}{s} = \begin{cases} (7.3 \pm 2.5) \times 10^{-11}, \text{ BBN} \\ (9.2 \pm 1.1) \times 10^{-11}, \text{ WMAP} \\ (8.59 \pm 0.11) \times 10^{-11}, \text{ Planck.} \end{cases} \tag{1.1}$$

This remarkable overlap in the estimates of the BAU from light element abundances (particularly deuterium) and baryon acoustic oscillations are shown in figures 1.1 and 1.2 respectively. Reproducing this estimate using particle physics makes one of the three pillars of the Standard Model of particle cosmology, the other two being inflation [5] and dark matter [6]. Like inflation and dark matter, it requires at least some additions to the Standard Model. Furthermore, like the other two pillars of the Standard Model of particle cosmology, there are a very large variety of models to explain this peculiar fact about our Universe. Two of the most elegant explanations

[1] For a recent attempted exception to this, albeit a fine-tuned one, see [1].

[2] Sometimes the baryon asymmetry is compared to the photon density, $n_B/n_\gamma \approx 7.04 Y_B$, rather than the entropy. However, during the early Universe many particles are expected to be in thermal equilibrium making Y_B more convenient.

doi:10.1088/978-0-7503-3571-3ch1

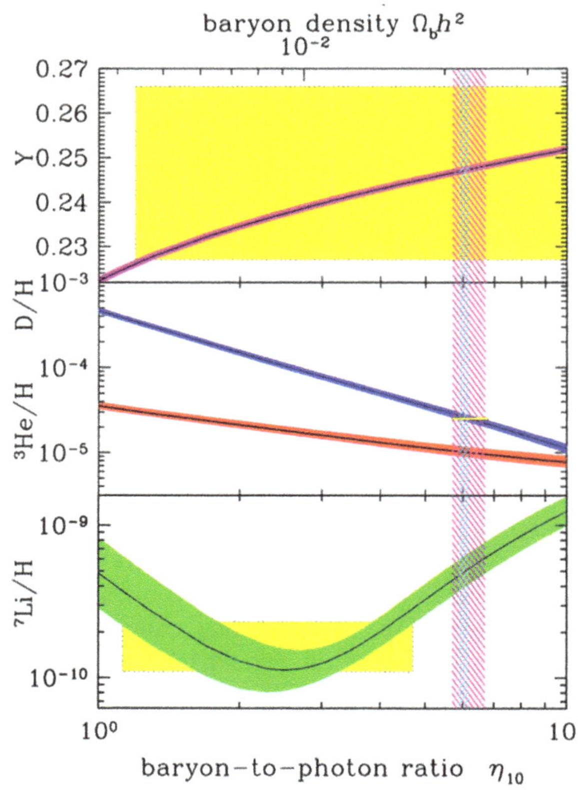

Figure 1.1. The abundances predicted by the Standard Model of Big Bang nucleosynthesis (BBN) [7] for ^{4}He, D, ^{3}He, and ^{7}Li. Here the bands show the range for the 95% confidence level and the boxes indicate the light element abundances—the smaller boxes show 2.75σ statistical errors; the larger boxes 2.75σ statistical and systematic errors. The wide band indicates the BBN concordance range, whereas the vertical narrow band indicates the baryon asymmetry measured via the CMB given at the 95% confidence level. Reproduced from [2], copyright (2004) with permission from Elsevier.

are leptogenesis [9] and the Affleck–Dine mechanism [10]. Unfortunately, both tend to be well out of reach of the particle colliders of today and of the foreseeable future.

The focus of this review will be electroweak baryogenesis, which is the term for any mechanism that produces the matter–antimatter asymmetry during the electroweak phase transition. Such a scenario requires physics beyond the Standard Model that must couple relatively strongly to Standard Model particles and have masses that are not too far above the weak scale. Therefore, unlike Affleck–Dine baryogenesis or leptogenesis, electroweak baryogenesis has the tantalizing prospect of being tested, at least indirectly, by weak and TeV scale searches at the large hadron collider (LHC).

Apart from testability, electroweak baryogenesis has the attractive feature that coincides the breaking of the symmetry between particles and anti-particles with the spontaneous breaking of the one symmetry we know to be broken—electroweak

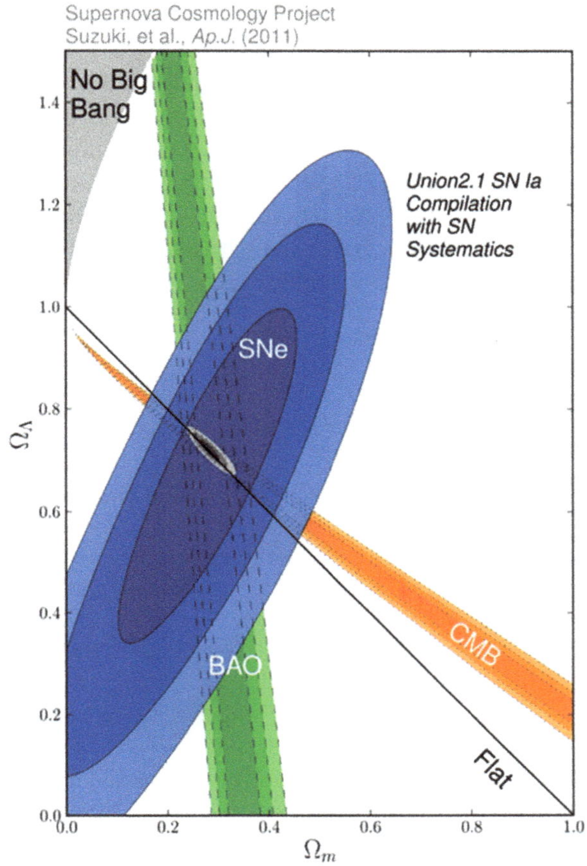

Figure 1.2. $\Omega_M - \Omega_\Lambda$ constraints due to CMB, baryon acoustic oscillations, and Supernova Cosmology Project Union 2.1 SN constraints including SN systematic errors. Reproduced from [8].

symmetry. Unfortunately, the literature on this exciting subject is somewhat opaque to newcomers. There are some very nice pedagogical introductions to small parts of the theoretical foundations of baryogenesis scattered throughout the literature if one digs hard enough[3]. However, the study of baryogenesis is arguably too decoupled from the rest of high energy physics given its promising phenomenological implications and the fundamental nature of the question it attempts to answer. Furthermore, the techniques one needs to learn to research in the field of electroweak baryogenesis have a large cross-over with other calculations in particle cosmology—including other models of producing the baryon asymmetry such as leptogenesis.

[3] For recent reviews see [11–14].

This primer will therefore give a mostly self-contained introduction to the field of *electroweak* baryogenesis, assuming the reader has a graduate level knowledge of particle physics, including dimensional regularization, some basic path integral techniques, the computation of amplitudes at tree and loop level, the Standard Model Lagrangian, and Big Bang cosmology, as well as a rudimentary knowledge of effective field theory and the minimal supersymmetric Standard Model (MSSM)[4], which we will use as a pedagogical tool where appropriate. There are of course many candidates for producing the BAU, but it is my hope that the theoretical foundations given in this book should make learning such mechanisms significantly more tractable.

References

[1] Krnjaic G 2017 Can the baryon asymmetry arise from initial conditions? *Phys. Rev.* D **96** 035041

[2] Eidelman S *et al* 2004 (Particle data group collaboration) Review of particle physics *Phys. Lett.* B **592** 1

[3] Spergel D N *et al* 2003 (WMAP Collaboration) First year Wilkinson Microwave Anisotropy Probe (WMAP) observations: determination of cosmological parameters *Astrophys. J. Suppl.* **148** A16

[4] Ade P A R *et al* 2014 Planck 2013 results. XVI. Cosmological parameters *Astron. Astrophys.* **571** A16

[5] Bassett B A, Tsujikawa S and Wands D 2006 Inflation dynamics and reheating *Rev. Mod. Phys.* **78** 2

[6] Feng J L 2010 Dark matter candidates from particle physics and methods of detection *Annu. Rev. Astron. Astrophys.* **48** 495–545

[7] Cyburt R H *et al* 2008 An update on the big bang nucleosynthesis prediction for ^{7}Li: the problem worsens *J. Cosmol. Astropart. Phys.* JCAP11(2008)012

[8] Suzuki N *et al* 2012 The hubble space telescope cluster supernova survey. V. Improving the dark-energy constraints above $z > 1$ and building an early-type-hosted supernova sample *ApJ* **746** 85

[9] Fukugita M and Yanagida T 1986 Baryogenesis without grand unification *Phys. Lett.* B **174** 1

[10] Affleck I and Dine M 1985 A new mechanism for baryogenesis *Nucl. Phys.* B **249** 361

[11] Morrissey D E and Ramsey-Musolf M J 2012 Electroweak baryogenesis *New J. Phys.* **14** 125003

[12] Cline J M 2012 Baryogenesis arXiv:0609145 (hep-ph)

[13] Riotto A and Trodden M 1999 Recent progress in baryogenesis *Annu. Rev. Nucl. Part. Sci.* **49** 46

[14] Trodden M 1999 Electroweak baryogenesis *Rev. Mod. Phys.* **71** 5

[15] Kuroda M 1999 Complete Lagrangian of MSSM arXiv:9902340 (hep-ph)

[16] Csaki C 1996 The minimal supersymmetric standard model (MSSM) *Mod. Phys. Lett.* A **11** 599–613

[4] If you lack knowledge of the MSSM see [15, 16] for an introduction.

IOP Publishing

Electroweak Baryogenesis (Second Edition)
An introduction
Graham White

Chapter 2

The Sakharov conditions

Perhaps the most convincing motivation for requiring an explanation for the BAU is the fact that inflation, a cornerstone of modern cosmology, will wash out any initial BAU. Before inflation was proposed, however, Sakharov proposed that any explanation of the BAU must satisfy three conditions, now famously known as the 'Sakharov conditions'. These conditions are:

- violation of baryon number conservation;
- violation of C and CP;
- departure from equilibrium.

The conditions are intuitively obvious as they basically say in crude terms that you need a process that allows you to create baryons, treat particles differently to anti-particles[1] and the future different to the past, and you need the process to change something in the Universe [1]. Even though the last condition is the most obvious it actually has a well-known loophole if CPT invariance is violated in a theory [2]. This can be seen from the proof of the third condition. Consider the equilibrium average of the baryon asymmetry B using standard thermal quantum mechanics[2],

[1] A mild subtlety is that C or CP violation separately does not achieve this. One requires both C and CP violation as a necessary condition for the rate of *total* baryon production to be different to total anti-baryon number.

[2] An astute reader might note that this argument implies that the baryon asymmetry is zero when the Universe returns to equilibrium. Indeed, processes that violate baryon number conservation do conspire to return the baryon number to zero on a time scale much larger than the age of the Universe. The key point of the argument however is that the baryon number of the Universe should be zero without some departure from equilibrium (or a violation of CPT).

$$\begin{aligned}\langle B\rangle &= \mathrm{Tr}[\mathrm{e}^{-\beta\mathcal{H}}B] \\ &= \mathrm{Tr}[(\mathrm{CPT})(\mathrm{CPT})^{-1}\mathrm{e}^{-\beta\mathcal{H}}B] \\ &= \mathrm{Tr}[\mathrm{e}^{-\beta\mathcal{H}}(\mathrm{CPT})^{-1}B(\mathrm{CPT})] \\ &= -\,\mathrm{Tr}[\mathrm{e}^{-\beta\mathcal{H}}B].\end{aligned} \tag{2.1}$$

The proof is quite elementary but does reveal that in asserting that baryon asymmetry production requires a departure from equilibrium you are implicitly assuming CPT invariance. This, however is a safe assumption in the vast majority of models including any this author is aware of that use the electroweak baryogenesis mechanism. Despite these conditions being intuitively obvious they are useful for the organization of this book, which is structured around these conditions. Remarkably one can, in principle, satisfy all three conditions using Standard Model particle content. However, the precise values of parameters within the Standard Model rule out it being sufficient to explain the BAU.

References

[1] Sakharov A D 1967 Violation of CP invariance, C asymmetry, and baryon asymmetry of the Universe *J. Exper. Theor. Phys.* **34** 392

[2] Bertolami O, Colladay D, Kostelecký V A and Potting R 1997 CPT violation and baryogenesis *Phys. Lett.* B **395** 178–83

IOP Publishing

Electroweak Baryogenesis (Second Edition)
An introduction
Graham White

Chapter 3

Baryon number violation in the Standard Model

The baryon number is a classically conserved quantity in the Standard Model as can be demonstrated by Noether's theorem, or even a cursory look at the available interactions. This intuition breaks down when one quantizes the Standard Model, due to so-called anomalous processes, which will be described in this chapter. That some fundamental symmetries may not survive the process of quantization is highly counter-intuitive. This is not just a feature of field theory. It was shown in [1, 2] that even standard quantum mechanics with certain potentials—specifically a delta function or a $1/r^2$ potential—have anomalous violations of conservation laws. The key anomaly of interest is the so-called chiral or axial anomaly. This anomaly, combined with the fact that the SU(2) gauge symmetry is a symmetry of left-handed particles only, will conspire to fulfil Sakharov's baryon number violation condition.

3.1 The axial anomaly

Let us warm up, however, by considering the simplest possible fermionic field theory, a single massless, free fermion field with a U(1) gauge symmetry. Such a theory has a global phase symmetry. Less obvious, is that such a model's Lagrangian is also invariant under the following transformation

$$\psi \to e^{i\theta\gamma_5}. \tag{3.1}$$

From these symmetries, we can of course use Neother's theorem to derive conserved currents. The Noether currents associated with these two symmetries are

$$J_\mu = \overline{\Psi}\gamma_\mu\psi \tag{3.2}$$

$$J_{5\mu} = \overline{\Psi}\gamma_5\gamma_\mu\psi, \tag{3.3}$$

doi:10.1088/978-0-7503-3571-3ch3

which implies Ward identities $\partial^\mu J_\mu = 0$ and $\partial^\mu J_{5\mu} = 0$. Obviously we expect the product of conserved currents to also be a conserved quantity. Therefore, we should not even need to bother to calculate the divergence of the following quantity:

$$-i\Gamma_{\mu\nu\lambda}(x_1, x_2, x_3) = \langle 0 \mid T\big(J_{5\mu}(x_3)J_\nu(x_1)J_\lambda(x_2)\big) \mid 0\rangle. \tag{3.4}$$

Amazingly this quantity is not zero. This remarkable result is known as the 'triangle anomaly' [3, 4] due to the fact that the Feynman diagram related to this amplitude has three vertices, as shown in figure 3.1. In the author's opinion the most effective way to learn the axial anomaly is through direct calculation. We present a review of what we believe is probably the simplest derivation in the literature using operator methods with dimensional regularization, followed by the Fujikawa method which is a path integral derivation.

3.1.1 Dimensional regularization

Let us calculate this example using the gauge invariant regularization scheme in dimensional regularization. A potential ambiguity that can arise in such a calculation is that there is no clear way to generalize the matrix γ_5 to 4-2ϵ dimensions, although various proposals exist [5]. We follow the approach of [6], which demonstrated that no knowledge of any of γ_5 properties is needed to calculate the triangle anomaly[1]. The axial current to one-loop order due to the triangle anomaly is given by

$$\langle 0 \mid J_5^\mu(x) \mid 0\rangle = \int \frac{d^D p}{(2\pi)^D}\frac{d^D k}{(2\pi)^D} e^{iq\cdot x} A_\nu(p) A_\lambda(k) \mathcal{M}^{\mu\nu\lambda}, \quad q = k + p. \tag{3.5}$$

The above amplitude relates to the two diagrams in figure 3.1, and the second diagram is obtained from the first by merely interchanging (p,ν) and (k,λ). Hence, it suffices to compute one diagram

$$\mathcal{M}^{\mu\nu\lambda}_{(1)}(k, p) = -ie^2 \int \frac{d^D l}{(2\pi)^D}\left\{\gamma^\mu\gamma_5 \frac{\not{l} - \not{k}}{(l-k)^2 + i\epsilon}\gamma^\lambda \frac{\not{l}}{l^2 + i\epsilon}\gamma^\nu \frac{\not{l} + \not{p}}{(l+p)^2 + i\epsilon}\right\}, \tag{3.6}$$

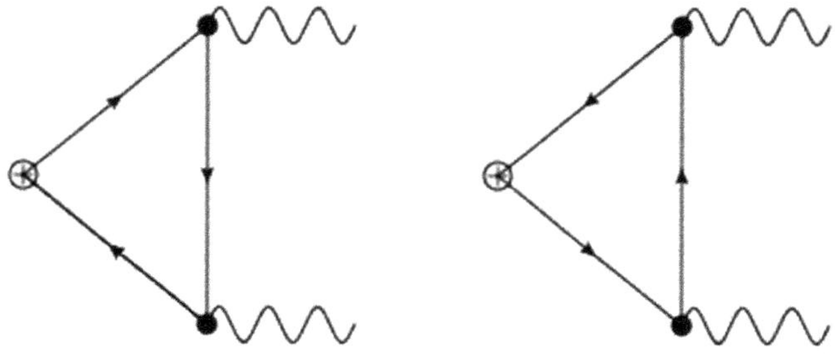

Figure 3.1. The famous triangle diagram responsible for anomalous quantum violations of classical chiral symmetry. The circle denotes a γ_5 on the vertex.

[1] Part of this calculation has some cross-over with the calculation in [7].

where the curly brackets denote a trace operation, and the photon momenta are on-shell. Although we introduced this section stating that we are considering a massless fermion model, we will use a mass to regularize the integrals and take the limit $m^2 \to 0$ at the end. Let us integrate over loop momentum and contract with q_μ. The result we can write as the sum of two terms

$$iq_\mu \mathcal{M}^{\mu\nu\lambda}_{(1)} = \mathcal{A}^{\nu\lambda}(k, p) + \mathcal{B}^{\nu\lambda}(k, p), \tag{3.7}$$

where

$$\begin{aligned} \mathcal{A}^{\nu\lambda}(k, p) = & \frac{ie^2}{32\pi^2} \int_0^1 dx \int_0^{1-x} dy \left[\frac{1}{\bar{\epsilon}} - \ln\left(\frac{\Delta}{\mu^2} \right) \right] \\ & \times \{ \not{q}\gamma_5\gamma^\alpha\gamma^\lambda\gamma_\alpha\gamma^\nu(x\not{k} + (1-y)\not{p}) - \not{q}\gamma_5(y\not{p} + (1-x)\not{k}) \\ & \times \gamma^\lambda\gamma^\alpha\gamma^\nu\gamma_\alpha - \not{q}\gamma_5\gamma^\alpha\gamma^\lambda(y\not{p} - x\not{k})\gamma^\nu\gamma_\alpha \} \end{aligned} \tag{3.8}$$

and

$$\begin{aligned} \mathcal{B}^{\nu\lambda}(k, p) = & -\frac{ie^2}{16\pi^2} \int_1^0 dx \int_{1-x}^0 dy \\ \times & \frac{\{ \not{q}\gamma_5((x-1)\not{k} - y\not{p})\gamma^\lambda(x\not{k} - y\not{p})\gamma^\nu(x\not{k} + (1-y)\not{p}) \}}{\Delta}. \end{aligned} \tag{3.9}$$

In the above we have defined the parameters

$$\Delta = -q^2xy + m^2, \quad \frac{1}{\bar{\epsilon}} = \frac{1}{\epsilon} - \gamma + \ln 4\pi. \tag{3.10}$$

The tensor $\mathcal{A}^{\nu\lambda}$ contains a divergent piece. If we employ the identities

$$\gamma^\alpha\gamma^\mu\gamma_\alpha = (-2+2\epsilon)\gamma^\mu, \quad \gamma^\alpha\gamma^\mu\gamma^\lambda\gamma^\nu\gamma_\alpha = -2\gamma^\nu\gamma^\lambda\gamma^\mu + 2\epsilon\gamma^\mu\gamma^\lambda\gamma^\nu \tag{3.11}$$

along with the cyclic property of traces, we can reduce this tensor to the form

$$\begin{aligned} \mathcal{A}^{\nu\lambda}_{\text{div.}} = \frac{ie^2}{96\pi^2\bar{\epsilon}} \Big[& 4p^\lambda\{\gamma^\nu\not{q}\gamma_5\} - 4k^\nu\{\gamma^\lambda\not{q}\gamma_5\} - 2p^\nu\{\gamma^\lambda\not{q}\gamma_5\} + 2k^\lambda\{\gamma^\nu\not{q}\gamma_5\} \\ & + 2g^{\nu\lambda}\{\not{k}\not{p}\gamma_5 - \not{p}\not{k}\gamma_5\} \Big]. \end{aligned} \tag{3.12}$$

So all traces with four Dirac matrices have already canceled and we are left with a series of terms that are zero in four dimensions. However, we need not make an assumption about the value of these traces. The term multiplying $1/\bar{\epsilon}$ is anti-symmetric under the interchange (p,ν) and (k,λ) so it cancels when the two diagrams are added, giving

$$\mathcal{A}^{\nu\lambda}_{\text{div.}}(k, p) + \mathcal{A}^{\lambda\nu}_{\text{div.}}(p, k) = 0. \tag{3.13}$$

The finite part of $\mathcal{A}^{\nu\lambda}$ comes from the $O(\epsilon)$ part of equation (3.11) and the $\ln \Delta$ part in equation (3.8). The result in the massless limit reads

$$\lim_{\epsilon\to 0} \quad [\mathcal{A}^{\nu\lambda}(k, p) + \mathcal{A}^{\lambda\nu}(p, k)]_{m=0} = \frac{e^2}{6\pi^2}\epsilon^{\lambda\nu\alpha\beta}p_\alpha k_\beta. \tag{3.14}$$

Note that the above tensor is infrared finite, so the inclusion of a mass regulator was a convenient calculation tool. Moving to $\mathcal{B}^{\nu\lambda}(k, p)$, we find that the potentially singular terms cancel identically so we can take the limit $D\to 4$ and write

$$\lim_{\epsilon\to 0} \quad [\mathcal{B}^{\nu\lambda}(k, p) + \mathcal{B}^{\lambda\nu}(p, k)]_{m=0} = \frac{e^2}{3\pi^2}\epsilon^{\lambda\nu\alpha\beta}p_\alpha k_\beta. \tag{3.15}$$

Putting everything together yields the well-known, but non-zero result

$$\langle 0 \,|\partial\cdot J_5 \,|\, 0\rangle = \frac{e^2}{2\pi^2}\epsilon^{\mu\nu\alpha\beta}\int \frac{d^Dp}{(2\pi)^D}\frac{d^Dk}{(2\pi)^D}e^{iq\cdot x}A_\nu(p)A_\lambda(k)\, p_\alpha k_\beta. \tag{3.16}$$

Thus we have found by explicit calculation the anomalous violation of chiral current conservation. We have chosen a particular way of performing the calculation that makes it maximally clear that the anomaly is not some ill effect of how we regulate our integrals. Indeed, the anomaly is experimentally demonstrated in predicting the correct rate for pion decay $\pi^0\to\gamma\gamma$. Furthermore, as remarked in the previous section, it has been shown to happen in ordinary quantum mechanics and the above calculation has been repeated with multiple different regularization schemes. Nonetheless we present one more derivation in the next subsection, since it naturally organizes the anomalies in the Standard Model such that it is straightforward to show that lepton and baryon currents have anomalous currents.

3.1.2 The Fujikawa method

Since the anomaly is independent of the regularization scheme we will calculate the anomalies using the path integral method with a hard cutoff regularization scheme. The regularization free nature of the anomaly is opaquer in this scheme, so I will assume that the reader understood the results of the previous section where things were more clear. The cost of this opaqueness will be justified by the simplicity of the calculation that follows. Let us begin by returning to the anomalous chiral current

$$J_\mu^5 = \overline{\Psi}\gamma_5\gamma_\mu\psi. \tag{3.17}$$

As in the previous section, the triangle diagram anomalously violates the classical conservation law derived through Noether's theorem. This time our fermions interact with Standard Model gauge bosons. Path integral techniques are far more efficient and systematic than trying to find all anomalous operators and hoping you have not missed any. The path integral's approach begins with simply adding the current to the Lagrangian [8, 9]

$$W[a_\mu, A_\lambda^c] = \int[d\psi][d\overline{\Psi}]e^{i\int d^4x\mathcal{L}_{QCD}-a_\mu J_5^\mu}. \tag{3.18}$$

We then vary W with respect to a_μ by an infinitesimal function $\partial_\mu \quad \beta$ in order to test whether W is invariant under such a change. Using the definition of functional derivatives

$$\begin{aligned}\delta[\ln W] &= \ln[W(a_\mu - \partial_\mu\beta, A_\mu^b)] - \ln[W(a_\mu, A_\mu^b)] \\ &= -\int d^4x \frac{\delta \ln W}{\delta a^\mu} \\ &= i\int d^4x \bar{J}_5^\mu \partial_\mu\beta \\ \delta[\ln W] &= -i\int d^4x \partial_\mu \bar{J}_5^\mu \beta(x).\end{aligned} \tag{3.19}$$

So if $\delta[\ln W] = 0$, $\partial_\mu J_5^\mu = 0$ and the chiral current is conserved. To test for an anomaly, one determines if the statement

$$W[a_\mu - \partial_\mu\beta, A_\lambda^c] = W[a_\mu, A_\lambda^c] \tag{3.20}$$

holds true. Let us begin by seeing if there is a change of variables that takes our primed Lagrangian back to the original form. That is

$$L'_{QCD}(\psi', \overline{\Psi}', A_\mu^a) \equiv L_{QCD}(\psi', \overline{\Psi}', A_\mu^a) - (\partial_\mu\beta)J_5^\mu \rightarrow L_{QCD}(\psi, \overline{\Psi}, A_\mu^a). \tag{3.21}$$

The appropriate set of transformations is

$$\psi' = (1 - i\beta\gamma_5)\psi \sim e^{-i\beta\gamma_5}\psi \tag{3.22}$$

$$\overline{\Psi}' = \overline{\Psi}(1 - i\beta\gamma_5) \sim \overline{\Psi}e^{-i\beta\gamma_5}. \tag{3.23}$$

This can be verified by direct substitution in which one ignores all terms $\mathcal{O}(\beta^2)$ and uses the anti-commutivity of the gamma matrices. The only other modification to make to our function W is the measure

$$[d\psi][d\overline{\Psi}] \rightarrow [d\psi'][d\overline{\Psi}'] \tag{3.24}$$

in which we will pick up a Jacobian determinant $\mathcal{J}$. We will later verify that $\mathcal{J}$ is independent of ψ and $\bar{\psi}$ so we can take it out of the integral

$$\begin{aligned}W\left[a_\mu - \partial_\mu\beta, A_\mu^a\right] &= \int [d\psi']\left[d\bar{\Psi}'\right] J e^{i\int d^4x L_{QCD} - a_\mu J_5^\mu} \\ &= \mathcal{J}\int [d\psi']\left[d\bar{\Psi}'\right] e^{i\int d^4x L_{QCD} - a_\mu J_5^\mu} \\ &= \mathcal{J}W\left[a_\mu, A_\mu^a\right].\end{aligned} \tag{3.25}$$

Therefore, we can write the variation of ln W exclusively in terms of the Jacobian

$$\begin{aligned}\delta[\ln W] &= \ln[W(a_\mu - \partial_\mu\beta, A_\mu^a)] - \ln[W(a_\mu, A_\mu^a)] \\ &= \ln[\mathcal{J}W(a_\mu, A_\mu^a)] - \ln[W(a_\mu, A_\mu^a)] \\ &= \ln \mathcal{J}.\end{aligned} \tag{3.26}$$

Our Jacobian is of Grassman valued functions so it is given by the inverse of the functional determinant of our transformations

$$\mathcal{J} = \text{Det}\,[\mathrm{e}^{i\beta\gamma_5}\mathrm{e}^{i\beta\gamma_5}]^{-1}. \tag{3.27}$$

We can use a simple identity, Det C = exp[Tr ln C], to write this determinant in terms of a trace

$$\mathcal{J} = \mathrm{e}^{-2i\mathrm{Tr}\beta\gamma_5}. \tag{3.28}$$

This determinant diverges, so we need to introduce a regulator. Let us first separate the divergent part of the trace

$$\mathrm{Tr}\beta\gamma_5 = \mathrm{Tr}' \int \mathrm{d}^4x \langle x \mid \beta\gamma_5\rangle, \tag{3.29}$$

where the primed trace is over internal degrees of freedom. To regularize the integral we rewrite it as

$$\mathrm{Tr}\,\beta\gamma_5 = \lim_{M\to\infty} \mathrm{Tr}' \int \mathrm{d}^4x \left\langle x \middle| \beta\gamma_5 \mathrm{e}^{-\left(\frac{\not{D}}{M}\right)^2} \middle| x \right\rangle. \tag{3.30}$$

The square in the regulator can be expanded

$$\left(\frac{\not{D}}{M}\right)^2 = \frac{1}{M^2}\left(\partial_\mu\partial^\mu + \frac{1}{4}g_3\lambda^a\sigma^{\mu\nu}F^a_{\mu\nu}\right) \equiv \frac{1}{M^2}(D^2 + g_3\sigma \cdot F). \tag{3.31}$$

Here the definition of $\sigma \cdot F$ is given implicitly. To calculate the trace of the regulator we introduce the complete set of states

$$\begin{aligned}\mathrm{Tr}\beta\gamma_5 &= \lim_{M\to\infty} \mathrm{Tr}' \int d^4x \frac{d^dp}{(2\pi)^{d/2}} \frac{d^dp'}{(2\pi)^{d/2}} \langle x|p\rangle\langle p|\beta\gamma_5 e^{-\left(\frac{\not{D}}{M}\right)^2}|p'\rangle\langle p'|x\rangle \\ &= \lim_{M\to\infty} \mathrm{Tr}' \int d^4x \frac{d^dp}{(2\pi)^d} \beta\gamma_5 e^{-(ip_\mu + D_\mu)^2 + g_3\sigma\cdot F)/M^2} \\ &= \lim_{M\to\infty} \mathrm{Tr}' \int d^4x \frac{d^dp}{(2\pi)^d} \beta\gamma_5 e^{p^2/M^2} e^{-(D^2 + g_3\sigma\cdot F + 2ip\cdot D)/M^2}.\end{aligned} \tag{3.32}$$

We will now Taylor expand the right-hand side up to $\mathcal{O}(1/M^4)$ keeping in mind that each factor of p^2 is of order $\mathcal{O}(M^2)$

$$
\begin{aligned}
\mathrm{Tr}\beta\gamma_5 = \lim_{M\to\infty}\mathrm{Tr}' \int \mathrm{d}^4x \frac{\mathrm{d}^d p}{(2\pi)^d}\beta\gamma_5 \mathrm{e}^{\frac{p^2}{M^2}}\Bigg[& 1 - \frac{1}{M^2}(D^2 + g_3\sigma\cdot F) \\
& + \frac{1}{2M^4}\left(\left\{D^2 + g_3\sigma\cdot F\right\}^2 - 4p\cdot D\right) + \frac{2}{3M^6} \\
& \times\ (p\cdot Dp\cdot D\{D^2 + g_3\sigma\cdot F\} + p\cdot D\{D^2 + g_3\sigma\cdot F\}p\cdot D) \\
& (+\ \{D^2 + g_3\sigma\cdot F\}p\cdot Dp\cdot D) + \frac{4}{M^8}(p\cdot D)^4\Bigg]
\end{aligned} \tag{3.33}
$$

$$
\begin{aligned}
&= \lim_{M\to\infty}\mathrm{Tr}' i \int \mathrm{d}^4x\beta\gamma_5 \frac{iM^4}{(4\pi)^2} \times \Bigg[1 - \frac{g_3\sigma\cdot F}{M^2} \\
&+ \frac{1}{M^4}\left(\frac{1}{2}(g_3\sigma\cdot F)^2 + \frac{1}{12}\left[D_\mu, D_\nu\right]^2 + \frac{1}{6}\left[D_\mu, \left[D_\mu, \sigma\cdot F\right]\right]\right)\Bigg].
\end{aligned}
$$

The only trace that survives is the one with a product of two σ. Recalling that

$$
\mathrm{Tr}\gamma_5\sigma^{\mu\nu}\sigma^{\alpha\beta} = -4i\epsilon^{\mu\nu\alpha\beta}, \tag{3.34}
$$

it is then easy to take the trace over the internal degrees of freedom to show that

$$
\mathcal{J} = \exp\left[-i \int \mathrm{d}^4x\beta\frac{3\alpha S}{8\pi}F^{\mu\nu}\widetilde{F}\mu\nu\right]. \tag{3.35}
$$

We therefore have

$$
\partial_\mu J_5^\mu = \frac{3\alpha_s}{8\pi}F^{\mu\nu}\widetilde{F}\mu\nu. \tag{3.36}
$$

This effective operator generates the amplitude given in equation (3.15) which was derived using operator techniques and dimensional regularization.

3.1.3 Baryon and lepton number violation

At this stage we have rederived the chiral anomaly using functional techniques and taking into account the gauge symmetries of the Standard Model. The SU(2) gauge bosons couple only to left-handed particles so the chiral anomaly interferes with lepton and baryon number conservation. Consider the classically conserved currents

$$
J^\mu_{L,\,R} = \frac{1}{2}\overline{\Psi}\gamma^\mu(1 \mp \gamma^5)\psi. \tag{3.37}
$$

Let us first couple the left-handed current to the Standard Model. Following the procedure as before we consider the term $W[a_\mu + \partial_\mu\beta, A_\mu, B_\mu, G_\mu]$ where $L = L_{\mathrm{SM}} + (a_\mu + \partial_\mu\beta)J^\mu_L$. This can be related to the term $W[a_\mu, A_\mu, B_\mu, G_\mu]$ by the transformations

$$
\psi' = (1 - i\beta\gamma_5)\psi \sim \mathrm{e}^{-i\beta\gamma_5}\psi \tag{3.38}
$$

$$
\overline{\Psi}' = \overline{\Psi}(1 - i\beta\gamma_5) \sim \overline{\Psi}\mathrm{e}^{-i\beta\gamma_5}. \tag{3.39}
$$

We then can use the method outlined before to derive an expression for $\partial_\mu J_L^\mu$ with associated regulator $\exp\{[(\partial_\mu + gA_\mu + g'B_\mu + g_s G_\mu)\gamma^\mu]^2/M^2\} \equiv \exp[\not{D}_L^2/M^2]$. Similarly we can couple the right-handed current to our Lagrangian. The transformation required to bring us back to our original W function is

$$\psi' = (1 + i\beta\gamma_5)\psi \sim \mathrm{e}^{+i\beta\gamma_5}\psi \tag{3.40}$$

$$\overline{\Psi}' = \overline{\Psi}(1 + i\beta\gamma_5) \sim \overline{\Psi}\mathrm{e}^{+i\beta\gamma_5}. \tag{3.41}$$

Our Jacobian will acquire an extra minus sign. This time the associated regulator is $\exp\{[(\partial_\mu + gA_\mu + g_s G_\mu)\gamma^\mu]^2/M^2\} \equiv \exp[\not{D}_R^2/M^2]$. Following the previous calculation one finds for quarks

$$\partial_\mu\,[J_L^\mu - J_R^\mu] \equiv \partial_\mu J_5^\mu = \frac{3\alpha_S}{8\pi} G^{\mu\nu}\widetilde{G}_{\mu\nu} \tag{3.42}$$

$$\partial_\mu J_L^\mu = \frac{3\alpha_W}{8\pi} B^{\mu\nu}\widetilde{B}_{\mu\nu}. \tag{3.43}$$

When one performs the calculation for leptons there is only the anomalous left-handed current which is equal to one third the above equation since one does not perform a trace over colour. Noting that the baryon number of a quark is 1/3 we can write the exact symmetry

$$\partial_\mu B^\mu = \partial_\mu L^\mu. \tag{3.44}$$

3.2 The Chern–Simons form, baryon number violation, and the winding number

We have now demonstrated that both lepton and baryon number are violated in the Standard Model through these strange anomalous currents. In this subsection we will try and find the sort of field configuration that violates baryon and lepton number. This is one part of the background theory which can quickly get unnecessarily formal, so we will be taking an approach to make things only as formal as is useful to understanding the subject at the level needed to do research in this field[2]. Let us first make some manipulations to our effective action. Recall that

[2] We do not follow any particular reference as they often lack details in the areas we need and are too formal or too detailed in areas orthogonal to this analysis. Nonetheless, some useful resources for more information are [10–12].

$$
\begin{aligned}
\partial_\mu L^\mu = \partial_\mu B^\mu &= \frac{g^2}{32\pi^2}\int \mathrm{d}^4x\epsilon^{\mu\nu\alpha\beta}\mathrm{Tr}[F_{\mu\nu}F_{\alpha\beta}] \\
&= \frac{g^2}{32\pi^2}\int \mathrm{d}^4x\epsilon^{\mu\nu\alpha\beta}\partial_\mu \mathrm{Tr}\Big[\partial_\mu A_\nu^a - \partial_\nu A_\mu^a + f^{abc}A_\mu^b A_\nu^c\Big] \\
&\qquad \times\Big[\partial_\alpha A_\beta - \partial_\beta A_\alpha^a + f^{ade}A_\alpha^d A_beta^e\Big] \\
&= \frac{g^2}{8\pi^2}\int \mathrm{d}^4x\epsilon^{\mu\nu\alpha\beta}\partial_\mu \mathrm{Tr}\Big[A_\nu\partial_\alpha A_\beta - \frac{2i}{3}A_\nu A_\alpha A_\beta\Big] \\
&\qquad + f^{abc}f^{ade}\epsilon^{\mu\nu\alpha\beta}\mathrm{Tr}\Big[[A_\mu^a, A\nu^b][A_\alpha^d A_\beta^e]\Big] \\
&= \frac{g^2}{8\pi^2}\int \mathrm{d}^4x\epsilon^{\mu\nu\alpha\beta}\partial_\mu \mathrm{Tr}\Big[A_\nu\partial_\alpha A_\beta - \frac{2i}{3}A_\nu A_\alpha A_\beta\Big] \\
&= \frac{g^2}{8\pi^2}\int_{S^3}\mathrm{Tr}\Big[A\wedge \mathrm{d}A + \frac{1}{3}A\wedge A\wedge A\Big].
\end{aligned}
\tag{3.45}
$$

Note we used the Jacobi identity in the above derivation. Also in the last line we have used the divergence theorem and wrote the result in a compact notation using wedge products. If you are unfamiliar with this notation rest assured we will use them sparingly. The integrand here is known as the 'Chern–Simons action' as it is the surface integral of the Chern–Simons form [13]. The field configurations that will be relevant are finite-energy solutions, so we require them to be well-behaved at infinity

$$
A_\mu \to ig\partial_\mu g^{-1} + \mathcal{O}\left(\frac{1}{r^2}\right). \tag{3.46}
$$

That is, we require that in some gauge the field vanishes at a rate faster than $1/r^2$ up to 'pure gauge' contributions. Field configurations that do not satisfy the finite-energy condition make no contribution to the functional integral so they can be ignored. We note that the integral of the Chern–Simons form is the volume integral of a total divergence. We can then use Gauss' divergence theorem to turn this into an integral over the surface at spatial infinity. If we enter the pure gauge part into equation (3.45) we obtain

$$
\begin{aligned}
\partial_\mu L^\mu &= \frac{g^2}{8\pi^2}\int \mathrm{d}S_\mu\epsilon^{\mu\nu\alpha\beta}\mathrm{Tr}\Big[i^2 g\partial_\nu g^{-1}(\partial_\alpha g\partial_\beta g^{-1}) - \frac{2i^4}{3}g\partial_\mu g^{-1}g\partial_\alpha g^{-1}g\partial_\beta g^{-1}\Big] \\
&= \frac{g^2}{8\pi^2}\int \mathrm{d}S_\mu\epsilon^{\mu\nu\alpha\beta}\mathrm{Tr}\Big[-g\partial_\nu g^{-1}(\partial_\alpha g(gg^{-1})\partial_\beta g^{-1}) \\
&\qquad - \frac{2}{3}g\partial_\mu g^{-1}g\partial_\alpha g^{-1}g\partial_\beta g^{-1}\Big] \\
&= \frac{g^2}{8\pi^2}\int \mathrm{d}S_\mu\epsilon^{\mu\nu\alpha\beta}\mathrm{Tr}\Big[g\partial_\mu g^{-1}g\partial_\alpha g^{-1}g\partial_\beta g^{-1} - \frac{2}{3}g\partial_\mu g^{-1}g\partial_\alpha g^{-1}g\partial_\beta g^{-1}\Big] \\
&= \frac{g^2}{24\pi^2}\int \mathrm{d}S_\mu\epsilon^{\mu\nu\alpha\beta}\mathrm{Tr}\Big[g\partial_\mu g^{-1}g\partial_\alpha g^{-1}g\partial_\beta g^{-1}\Big],
\end{aligned}
\tag{3.47}
$$

where in the second line we used the identity

$$\partial_x 1 = 0 = \partial_x(gg^{-1}) = (\partial_x g)g^{-1} + g(\partial_x g^{-1}). \tag{3.48}$$

The form of this is very interesting. Consider the volume element for the internal group

$$\mathrm{d}\Omega_V = \frac{1}{24\pi^2}\int \mathrm{Tr}[(\mathrm{d}gg^{-1}) \wedge (\mathrm{d}gg^{-1}) \wedge (\mathrm{d}gg^{-1})]. \tag{3.49}$$

So the particle current divergence is proportional to the volume integral over the internal group pulled back onto Euclidean space–time. In other words it is proportional to the number of times the field winds around the group at the three-dimensional surface at spatial infinity. If the integral of the Chern–Simons form is non-zero we therefore denote it with a 'non-zero' Chern–Simons number due to the connection with the winding number.

3.3 Winding number and non-abelian gauge groups

The kind of field configuration that violates baryon number conservation is one with a non-zero 'Chern–Simons number', which we argued in the previous section is satisfied if the winding number is non-zero. In this section we will make things a little more concrete. The winding number is a concept from topology. As we only need surface level understanding of the concept we will only give a few examples to build intuition toward the subject. First consider the case where the internal space of a gauge field is a simple circle. Now consider the following mapping:

$$g^{\nu}(\theta) = \mathrm{e}^{(i\nu\theta)}. \tag{3.50}$$

The winding number is defined as how often this function winds around the circle. It is easy to show that the following equation is the winding number:

$$\frac{i}{2\pi}\int_0^{2\pi} \mathrm{d}\theta \mathrm{d}g^{-1}/\mathrm{d}\theta. \tag{3.51}$$

What is obvious also from explicit calculation is that the product of two mappings gives a winding number that is the sum of the winding numbers of each mapping. That is, if the winding number of one function, $f(\theta)$ is ν_1 and the winding number of $g(\theta)$ is ν_2, then the winding number of $f(\theta)g(\theta)$ is $\nu_1 + \nu_2$. This just follows from the properties of exponential functions. However, let us prove this in a way that may seem needlessly complicated for this simple example, but will be useful for the more complicated case we are interested in. First we consider a small change in the winding number

$$\delta g = i(\delta\lambda). \tag{3.52}$$

It is easy to show that

$$\delta\nu = 0,$$

$$\begin{aligned}\nu + \delta\nu &= \frac{i}{2\pi}\int_{2\pi}^{0} \mathrm{d}\theta \quad (g + \delta g)\frac{\mathrm{d}(g^{-1} - \delta g)}{\mathrm{d}\theta}\mathrm{d}\theta \\ &= \nu + \frac{1}{2\pi}\int_{2\pi}^{0}\left(\delta g\frac{\mathrm{d}g^{-1}}{\mathrm{d}\theta} - i\frac{\mathrm{d}\delta\lambda g^{-1}}{\mathrm{d}\theta}\right) \\ &= \nu + \frac{1}{2\pi}\int_{2\pi}^{0}\mathrm{d}\theta\frac{\mathrm{d}\delta\lambda}{\mathrm{d}\theta} \\ &= \nu.\end{aligned} \tag{3.53}$$

Thus we have shown that the winding number is invariant under continuous transformations. So we see that the winding number of the product of two functions is just the sum of the two winding numbers. Now consider the case where the gauge field has a local SU(2) symmetry. In the previous section we claimed that the winding number for this group is given by

$$\nu = \frac{1}{24\pi^2}\int \mathrm{d}S_\nu \epsilon^{\nu\mu\alpha\beta}\mathrm{Tr}[g\partial_\mu g^{-1} g\partial_\alpha g^{-1} g\partial_\beta g^{-1}], \tag{3.54}$$

which we will prove to be the winding number in a way completely analogous to the case of a circle. It is straightforward by direct substitution to take a configuration with winding number of unity, $g = \sigma\cdot\mathbf{x}/r$ (where of course $\sigma^4 = 1_2$), and show that it indeed gives one when inserted into the above equation. So we can prove it is the winding numbing by demonstrating that winding number of two maps is the sum of their individual winding numbers using the same techniques we used for the circle. Once again we begin with considering a small change in the winding number. Recall the identity that

$$(\partial_x g)g^{-1} = -g(\partial_x g^{-1}), \tag{3.55}$$

which we can use to write

$$\delta\nu = \int \mathrm{d}\sigma_\nu \epsilon^{\nu\mu\alpha\beta}\mathrm{Tr}[\partial_\mu g^{-1}\partial_\alpha g \partial_\beta \delta\lambda^a(x)\tau^a]. \tag{3.56}$$

This vanishes upon integration by parts. It is then trivial to see that our expression for ν is indeed the winding number as one can then continuously deform our expression for g such that it is equal to one for the upper part of the hypersphere, and deform another mapping of winding number one such that it is equal to one for the lower part of the hypersphere. The winding number of the product of these two mappings is then 2. A similar process can produce a mapping g_ν with winding number ν. So what does this mean? If we consider the SU(2) gauge fields in four-dimensional Euclidean space and track their value as we walk around spatial infinity, if the field winds around its internal group at least once in this walk then baryon and lepton number is violated. So to understand the process of baryon number violation we will then seek to understand these topologically interesting processes. Specifically we will find in the following sections that we are interested in lumps of field, from the perspective of four-dimensional Euclidean space, called instantons, and unstable temporally localized lumps called sphalerons.

3.4 Solitons and instantons

In the previous sections we demonstrated that field configurations that violate baryon or lepton number must be topologically non-trivial—that is they must have non-zero Chern–Simons number. In this section we will demonstrate that topologically interesting field configurations generically are associated with quasi-particles.

Let us begin with the simplest example. Consider a simple scalar field in 1 + 1 dimensions with degenerate minima. For simplicity and clarity we will set a bunch of parameters in this theory to 1 to avoid them cluttering our equations as we only wish to discuss some qualitative features of the theory. The 'uncluttered' version of the Lagrangian for 1 + 1-dimensional scalar field theory is

$$\mathcal{L} = \frac{1}{2}\left(\frac{\mathrm{d}\phi}{\mathrm{d}t}\right)^2 - \frac{1}{2}\left(\frac{\mathrm{d}\phi}{\mathrm{d}x}\right)^2 - \frac{1}{4}(\phi^2 - 1)^2. \tag{3.57}$$

Now consider finite-energy solutions to the action that are stationary in time (that is, we set the $\mathrm{d}\phi/\mathrm{d}t$ term to zero). Without putting pen to paper we can already make remarks about the types of solutions we will find. The key to this is knowing that if the solution is finite then the field must take the value of one of the minima at plus and minus spatial infinity. This gives us two types of solution, a field that has the same minima at $\pm\infty$ and a field that goes from one minimum at $-\infty$ to another at $+\infty$. The second class of solutions is topologically non-trivial. Consider a specific example of the second type of solution to the classical equations of motion

$$\phi(x) = \pm\tanh[(x - x_0)/\sqrt{2}]. \tag{3.58}$$

The energy profile of this solution is a highly localized lump of field. This can be seen if we write the energy density of the solution

$$\epsilon(x) = \frac{1}{2} \quad \mathrm{sech}^2[(x - x_0)/\sqrt{2}] \tag{3.59}$$

which we depict in figure 3.2.This is known as a classical lump or a soliton. It is a quasi-particle whose existence derives from a degeneracy in the ground state. Although the solution given above is stationary, we can easily boost to a moving

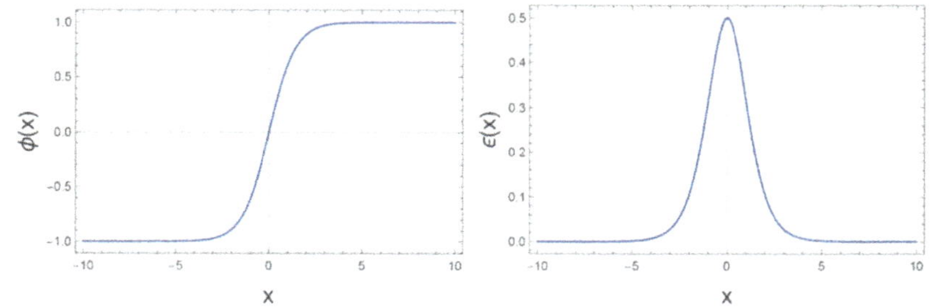

Figure 3.2. Depiction of a soliton in a single spatial dimension for the simple case of two degenerate minima at ±1. The energy profile is highly localized which justifies its status as a quasi-particle.

frame to obtain moving soliton solutions. Recalling that under a Lorentz transformation, $\phi(x)\rightarrow\phi(\Lambda x)$, we can write the following class of moving solitons

$$\phi(x, t) = \tanh\left[\frac{1}{\sqrt{2}}\left(\frac{(x - x_0) - vt}{\sqrt{1 - u^2}}\right)\right]. \tag{3.60}$$

In the Standard Model we have a very large degeneracy in the ground state since one can have different winding numbers. Field configurations that are topologically interesting in an analogous way to the soliton described above will also look like localized lumps in the field except in all four Euclidean dimensions. These lumps are known as instantons since they are localized in time. If you reverse the Wick rotation, returning to 3 + 1 dimensions, these instantons look like tunneling processes [14]. Having more dimensions and a more complicated degeneracy means we cannot immediately write down the instanton solution as we did with the soliton. In the case where we have SU(2) gauge symmetry, it helps to first show that the instanton is a solution to the classical equations of motion—it is a local minimum of the action. First take the field strength tensor and make use of the trivial identity

$$\begin{aligned}\int d^4x \,\mathrm{Tr}\, F_{\mu\nu}F^{\mu\nu} &= \int d^4x \,\mathrm{Tr}\left[\sqrt{F_{\mu\nu}F^{\mu\nu}\widetilde{F}_{\alpha\beta}\widetilde{F}^{\alpha\beta}}\right] \\ &\geqslant \int d^4x \,\mathrm{Tr}\left[\left|F_{\mu\nu}\widetilde{F}^{\mu\nu}\right|\right].\end{aligned} \tag{3.61}$$

This means that

$$\int d^4x \mathrm{Tr} F_{\mu\nu}F^{\mu\nu} \geqslant \frac{8\pi^2}{g^2} \mid \nu \mid \tag{3.62}$$

with the equality being satisfied for $F = \widetilde{F}$. To construct an ansatz for an instanton we have the conditions that the winding number is 1, the configuration is finite-energy and the solution is a local extrema of the action. For simplicity let us ignore the gauge coupling constant by setting it to 1 and write the ansatz

$$A_\mu = if(r^2)g(x)\partial_\mu g^{-1}(x), \tag{3.63}$$

where f is a radial function that goes to 1 as $r\rightarrow\infty$ faster than $1/r^2$. We can substitute this ansatz into the classical equations of motion (or equivalently the equation $F = \widetilde{F}$) to find

$$f(r^2)^2 - f(r^2) = -r^2 f'(r^2), \tag{3.64}$$

which is satisfied for

$$f(r^2) = \frac{r^2}{r^2 + \rho^2}. \tag{3.65}$$

The instanton is then [15, 16]

$$A_\mu(x) = g^{-1}(x-a)[\partial_\mu g(x-a)]\frac{|(x-a)|^2}{|(x-a)|^2+\rho^2}$$
$$\phi(x) = 1 + \frac{\rho^2}{(x-a)^2}. \qquad (3.66)$$

That $A_\mu(x)$ winds around the group once is clear from the fact that it is a radial function times the pure gauge form for the $\nu = 1$ gauge configuration. This can also be seen by explicitly putting the above solution into the Chern–Simons form (3.47). The above instanton tunnels between two neighboring vacua and in the process produces three leptons and three baryons. If all particles are Standard Model this means nine quarks and three leptons. Calculating the energy density of this solution once again shows a clump of field highly localized in space and time, as can been seen in figure 3.3.

The amplitude for an instanton rate at zero temperature can be found by simply inserting the instanton solution into the path integral and making a saddle-point approximation. Doing so one finds that the instanton amplitude is proportional to $\exp[-8\pi^2/g^2] \sim 10^{-173}$ [14] which is an incredible suppression! So at zero temperature it cannot possibly be important to explain the baryon asymmetry. Finally, let us see how the ground state of the Standard Model can be thought of as a periodic potential [17–19]. Let an eigenstate of definite winding number be denoted by $|n\rangle$. The conjugate variable to n will then be a phase variable. An eigenstate of phase can be written as

$$|\theta\rangle = \sum_{n=-\infty}^{\infty} e^{in\theta}|n\rangle. \qquad (3.67)$$

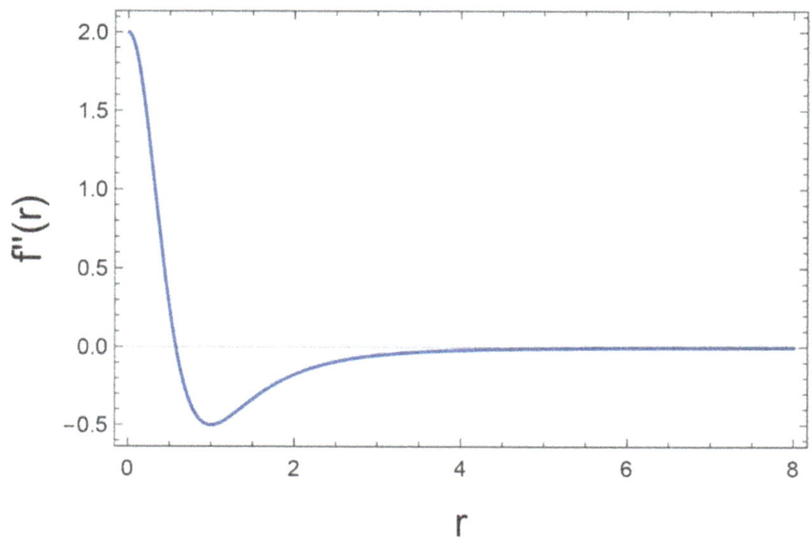

Figure 3.3. The energy profile of an instanton as a function of r^2 is also highly localized. This is the motivation behind thinking of the instanton as a quasi-particle in Euclidean space. Here the parameter $\rho = 1$.

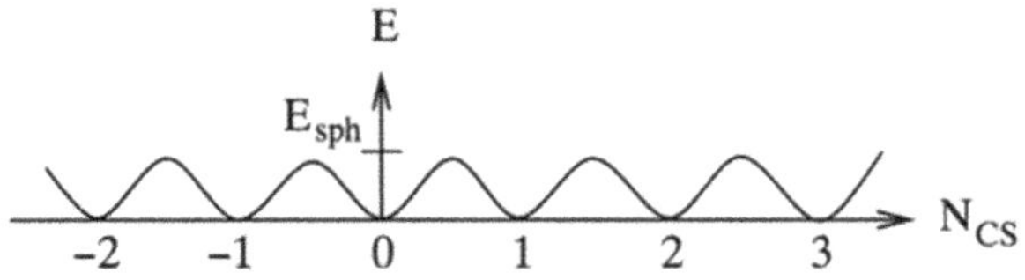

Figure 3.4. Standard Model ground state is actually periodic with a potential height of E_{sph}. Each degenerate minimum corresponds to a different Chern–Simons number. A shift to the right from one minimum to another is the instanton, whereas the local maxima are the sphalerons. Each degenerate ground state has a different Chern–Simons number, N_{CS}. Instantons move one to the right from one ground state to another with a higher Chern–Simons number, whereas anti-instantons move one to the left. Figure taken from [22].

Let us consider the term $\langle\theta'|\exp[-HT]|\theta\rangle$. We of course have calculated the term $\langle n+1 \mid \exp\ [-HT]|\ n\rangle - \langle n+1 \mid HT \mid n\rangle \approx AT\ \ \exp[-8\pi^2/g^2]$. We can make the approximation where we ignore multiple tunneling events and consider only single tunneling transitions. Defining $E(\theta)\delta(\theta-\theta') = \langle\theta'|H|\theta\rangle$ we have

$$E(\theta) = E_0 - 2A\exp\left[\frac{-8\pi^2}{g^2}\right]\cos\theta. \tag{3.68}$$

The ground state can be approximated as a periodic potential with a potential height proportional to the tunneling amplitude, as depicted in figure 3.4.

Let us conclude the section by summarizing what we have learned so far about baryon number violation in the Standard Model:

- Classical conservation laws can be anomalously broken.
- Triangle anomalies in the Standard Model result in an anomalous violation of both baryon and lepton number conservation.
- The field configurations that anomalously violate baryon/lepton number are ones with a non-zero Chern–Simons number. This means they wrap around the internal space at the surface at spatial infinity in Euclidean space.
- Fields with integer winding number at infinity are solutions to the classical equations of motion.
- Fields with integer winding number look like localized lumps in Euclidean space.

3.5 The sphaleron

Let us now try and bootstrap another topologically interesting quasi-particle that is a good candidate for baryon number violation. Specifically, the quasi-particle we will discuss is the sphaleron whose etymology is from the Greek word that means to 'go down', because it is an unstable particle [20, 21]. In the previous section we presented the instanton with a winding and Chern–Simons number of 1. It turns out that the instanton is less important than the sphaleron for our purposes because sphaleron rates become unsuppressed at high temperatures. As such we give it far closer attention. Physically, the instanton corresponds to tunneling through a barrier, whereas a sphaleron corresponds to classically passing over the barrier. A sphaleron is an unstable particle corresponding to an instantaneous moment in time and is a solution to the classical equations of motion. It is due to its importance in the logical framework of this field that we give it such deep attention. It should be noted that we only focus on the lowest energy electroweak sphaleron before touching on the SU(3) sphaleron, since the lowest energy sphaleron will dominate. Of course we note that there is no formal proof that we are aware of that the sphaleron we present is indeed the lowest energy but it is widely believed to be so. To begin our bootstrap let us start with insisting that the Higgs field, which is an $SU(2)_L$ doublet, is well-behaved at infinity. Well-behaved in this case means the covariant derivative must vanish

$$D_\mu\phi \to 0 \text{ as } r \to \infty. \tag{3.69}$$

This gives the condition

$$\partial_\mu\phi \Big|_\infty = igA_\mu^a\tau^a \Big|_\infty \phi \big|_\infty. \tag{3.70}$$

In this case we keep the gauge coupling constant as we will eventually calculate the sphaleron rate. We want $\partial\phi/\partial t = 0$ and we want a topologically non-trivial solution. So let us set $A_0 \,|_\infty$ to zero along with $\mathrm{d}\phi/\mathrm{d}t \,|_\infty$ and map the space–time surface onto the internal space by setting A_i^∞ to be proportional to the right invariant one form—also called the Maurer–Cartan form

$$\phi^\infty = \Omega\begin{pmatrix} 0 \\ v/\sqrt{2} \end{pmatrix} \tag{3.71}$$

and

$$A_i^\infty = \frac{-i}{g}(\partial_i\Omega)\Omega^{-1}. \tag{3.72}$$

We will use Ω instead of g here and throughout for an element of the gauge group to avoid confusion with the coupling constant. The Hermitian conjugate of the above is a solution. Both of these solutions can be verified by direct substitution.

This is the first requirement of a sphaleron: that it reduce to the Maurer–Cartan form at infinity in order to be a finite-energy solution. Fields with integer winding number are known to be (multi-)instantons. The only other topologically interesting cases that can potentially violate baryon number are the cases where we have a non-integer and non-zero Chern–Simons number. It will turn out that sphalerons have a Chern–Simons number of 1/2. However, the Chern–Simons number was shown in the previous section to be related to the winding number, which is forced to be an integer. So it is something of a surprise that a non-integer Chern–Simons number is possible given its close relationship to the winding number. Indeed the Chern–Simons number of any *vacuum* state must be an integer so a non-integer Chern–Simons number means it must continuously deform from one vacuum state to another.

To complete our bootstrap we explicitly make use of the Pauli matrices and make our fields cover the internal group as we cover the spatial sphere at $r = \infty$ while passing through zero somewhere in space via use of a radial function. Therefore, a suitable ansatz is

$$\phi^{\infty} = h(r)\Omega\begin{pmatrix} 0 \\ v/\sqrt{2} \end{pmatrix} \tag{3.73}$$

and

$$A_i^{\infty} = \frac{-if(r)}{g}(\partial_i\Omega)\Omega^{-1} \tag{3.74}$$

with

$$\Omega = \vec{\sigma} \cdot \frac{\vec{x}}{r}. \tag{3.75}$$

Also, $f \equiv f(r)$ and $h(r)$ are radial functions chosen to solve the equations of motion subject to the boundary conditions $h(0) = f(0) = 0$ and $h(\infty) = f(\infty) = 1$. Let us now complete the motivation for studying closely this quasi-particle over the instanton, by demonstrating that these processes become unsuppressed when electroweak symmetry is restored at high temperature. We apologize to the reader for slightly skipping ahead in the flow of our logical argument by using a small amount of finite temperature field theory from the later parts of this primer. The reader should regard the form of the next few equations as divinely given and then return to this section after reading the later sections on finite temperature quantum field theory (QFT). The sphaleron rate can be found by substituting the above solution into the finite temperature version of the path integral. The high temperature sphaleron rate is given by

$$\Gamma_{\rm sph} = \frac{\omega_-}{2\pi}(\mathcal{N}V)_{\rm rot}(\mathcal{N}V)_{\rm tr}\kappa e^{\Delta E_{\rm sph}/T}, \tag{3.76}$$

where $\Delta E_{\rm sph}/T = S_3(\phi_{\rm sph})$, where

$$S_3[\phi_{\rm sph}] = \frac{1}{T}\int d^3x \mid D_i\phi \mid^2 + \frac{1}{4}W_{ij}^a W_{ij}^a + V(\phi). \tag{3.77}$$

Here W_{ij} are the field strength tensors for A_i, $W_{ij} \equiv \partial_i A_j - \partial_j A_i$. With some effort we can reduce this to

$$S_3[\phi_{\rm sph}] \equiv \frac{\Delta E_{\rm sph}}{T} = \frac{4\pi v(T)}{gT}B\left(\frac{\lambda}{g^2}\right) \tag{3.78}$$

with

$$\begin{aligned} B\left(\frac{\lambda}{g^2}\right) = \int_{\infty}^{0} dr\Bigg[4\left(\frac{df}{dr}\right)^2 + \frac{8}{r^2}f^2(1-f)^2 \\ + \frac{r^2}{2}\left(\frac{dh}{dr}\right)^2 h^2(1-f)^2 + \frac{1}{4}\frac{\lambda}{g^2}r^2(1-h^2)^2\Bigg], \end{aligned} \tag{3.79}$$

where $r \equiv gv(T)|\, x\,|^2$. Note that when $v(T) = 0$ the rate is no longer exponentially suppressed. We will show in later chapters that electroweak symmetry is indeed restored at very high temperatures, meaning that at some stage in our cosmic history sphalerons, and therefore baryon number violation, were unsuppressed! This is the key feature of sphalerons which makes them attractive as a source of baryon number violation at very high temperature when electroweak symmetry is restored. There is one remaining thing we have to show to conclude our discussion of sphalerons in terms of showing that we indeed have a non-zero Chern–Simons number. Recall that the Chern–Simons form is

$$\mathrm{Tr}\left[W \wedge \mathrm{d}W + \frac{2}{3} W \wedge W \wedge W\right] \tag{3.80}$$

where for $A_0 = 0$, the above can be written

$$\begin{aligned} \mathrm{Tr}[W \wedge \mathrm{d}W] &= \epsilon_{ijk} W_i\left(\partial_j W_k - \partial_j W_j\right) = -\epsilon_{ijk}\left(\partial_i W_j - \partial_j W_i\right) W_k \\ \mathrm{Tr}[W \wedge W \wedge W] &= - \quad \epsilon_{ijk}\epsilon_{abc} W_i^a W_j^b W_k^c \end{aligned} \tag{3.81}$$

and the Chern–Simons number for the sphaleron is

$$Q(\text{sphaleron}) = \frac{g^2}{32\pi^2} \int_{t=t_0} \mathrm{d}^3x K^0, \tag{3.82}$$

where K° is the Chern–Simons form. Recall our field configuration,

$$W_i^a \sigma^a = -\frac{2if}{g} \Omega^{-1} \mathrm{d}\Omega \tag{3.83}$$

where Ω is given above and $\mathrm{d}\Omega$ is the external derivative $\frac{\partial \Omega}{\partial x_i} \mathrm{d}x_i$. Remarkably the form of f will not affect the Chern–Simons number as long as the boundary conditions given above are met. It is easy to see that the above equation reduces to

$$W_i^a \mathrm{d}x_i = -\frac{2f}{g} \epsilon_{iab} x_b. \tag{3.84}$$

We cannot calculate the Chern–Simons number in this form as the derivation of the Chern–Simons number assumes that our field goes to zero faster than $1/r$. Let us therefore perform a gauge transformation $U = \exp\left[\frac{i}{2}\Theta(r)\sigma \cdot \frac{x}{r}\right]$, where $\Theta(r)$ is a radial function that goes to zero at the spatial origin. Recall that a gauge transformation for non-abelian fields has the form

$$gW_i^a \sigma_a \to U^\dagger\left(\frac{-2f}{r^2} \epsilon_{aik} x_k \sigma_a\right) U + i(\partial_i U^{-1}) U. \tag{3.85}$$

It is straightforward to show that in this gauge the sphaleron solution looks like

$$gW_i^a = A\epsilon_{iab} x_b + B(\delta_{ia} r^2 - x_i x_a) + \Theta' \frac{x_i x_a}{r^2} \equiv \mathcal{A}_{ai} + \mathcal{B}_{ai} + \mathcal{C}_{ai}, \tag{3.86}$$

where the prime denotes a radial derivative and

$$
\begin{aligned}
A \equiv A(r) &= \frac{1}{r^2}[(1 - 2f(r))\cos(\Theta) - 1] \text{ and} \\
B \equiv B(r) &= \frac{1}{r^3}[(1 - 2f(r))\sin(\Theta)].
\end{aligned}
\tag{3.87}
$$

Such abbreviations may be disorientating but we will continue to use them throughout because it becomes far too cumbersome to write out every term explicitly. The sphaleron now has the correct asymptotic behavior. Since ${W^a}_i$ is spherically symmetric the integral over angular variables in the calculation is trivial allowing us to rewrite the Chern–Simons number of the sphaleron as

$$
Q(\text{sphaleron}) = \frac{g^2}{8\pi}\int_{t=t_0} r^2 \mathrm{d}r K^0. \tag{3.88}
$$

Using above equations and verifying this with equation (3.45) we can write

$$
K^0 = -\left(\epsilon_{ijk}F_{ij}^a W_k^a + \frac{2g}{3}\epsilon_{ijk}\epsilon_{abc}W_i^a W_j^b W_k^c\right). \tag{3.89}
$$

Let us first calculate the first term in the Chern–Simons density. Due to our gauge transformation it consists of three terms

$$
\epsilon_{ijk}F_{ij}^a W_k^a \equiv (A_{ak} + B_{ak} + C_{ak})W_k^a, \tag{3.90}
$$

which in turn are defined by the three terms we defined before $\mathcal{A}_{ia}$, $\mathcal{B}_{ia}$ and $\mathcal{C}_{ia}$ as follows

$$
A_{ak} \equiv \epsilon_{ijk}[\partial_i \mathcal{A}_{ja} - \partial_j \mathcal{A}_{ia}] \text{ etc.} \tag{3.91}
$$

With a bit of effort one can simplify the expressions for (A_{ak}, B_{ak}, C_{ak}), reducing them down to the simple form

$$
\begin{aligned}
gA_{ak} &= \frac{2}{r}[r^2\delta_{ka} - x_k x_a]A' + 4A\delta_{ka} \\
gB_{ak} &= 2\epsilon_{iak}x_i[B'r + 3B]
\end{aligned}
\tag{3.92}
$$

$$
gC_{ak} = -2\epsilon_{iak}x^i\Theta'. \tag{3.93}
$$

We now contract the above with ${W^a}_k$ which gives a sum of nine terms. It is easy to show that these terms are

$$
\begin{array}{lll}
g^2 A_{ka}\mathcal{A}_{ka} = 0 & g^2 A_{ka}\mathcal{B}_{ka} = 4BA'r^3 + 8ABr^2 & g^2 A_{ka}\mathcal{C}_{ka} = 4A\Theta' \\
g^2 B_{ka}\mathcal{A}_{ka} = -4[B'r + 3B]Ar^2 & g^2 B_{ka}\mathcal{B}_{ka} = 0 & g^2 B_{ka}\mathcal{C}_{ka} = 0 \\
g^2 C_{ka}\mathcal{A}_{ka} = 4A\Theta' r^2 & g^2 C_{ka}\mathcal{B}_{ka} = 0 & g^2 C_{ka}\mathcal{C}_{ka} = 0.
\end{array}
\tag{3.94}
$$

So we have, in summary, the first term in the Chern–Simons density calculated as

$$\frac{g^2r^2}{8\pi}\epsilon_{ijk}F^a_{ij}W^a_k = \frac{1}{2\pi}BA'r^5 + \frac{1}{\pi}ABr^4 + \frac{1}{2\pi}r^4A[B'r + 3B] + \frac{1}{\pi}A\Theta'r^4. \quad (3.95)$$

Next we have the term $g^2\epsilon_{ijk}\epsilon_{abc}W^a_iW^b_jW^c_k$. Let us first calculate the product of two sphalerons contracted with the permutation symbols

$$\epsilon_{ijk}\epsilon_{abc}W^a_iW^b_j = \epsilon_{abc}\epsilon_{ijk}[\mathcal{A}_{ai} + \mathcal{B}_{ai} + \mathcal{C}_{ai}][\mathcal{A}_{bj} + \mathcal{B}_{bj} + \mathcal{C}_{bj}]. \quad (3.96)$$

It is useful to organize the terms into a matrix

$$\begin{aligned} &g^2\epsilon_{ijk}\epsilon_{abc}\begin{pmatrix} \mathcal{A}_{ai}\mathcal{A}_{bj} & \mathcal{A}_{ai}\mathcal{B}_{bj} & \mathcal{A}_{ai}\mathcal{C}_{bj} \\ \mathcal{B}_{ai}\mathcal{A}_{bj} & \mathcal{B}_{ai}\mathcal{B}_{bj} & \mathcal{B}_{ai}\mathcal{C}_{bj} \\ \mathcal{C}_{ai}\mathcal{A}_{bj} & \mathcal{C}_{ai}\mathcal{B}_{bj} & \mathcal{C}_{ai}\mathcal{C}_{bj} \end{pmatrix} \\ &= \begin{pmatrix} 2x_kx_cA^2 & 0 & -\Theta'A\epsilon_{kbc}x_b \\ 0 & 2x_kx_cB^2r^2 & B\Theta'(\delta_{kc}r^2 - x_kx_c) \\ -\Theta'A\epsilon_{kbc}x_b & B\Theta'(\delta_{kc}r^2 - x_kx_c) & 0 \end{pmatrix} \\ &\equiv M. \end{aligned} \quad (3.97)$$

Contracting the last sphaleron gives the following

$$\begin{aligned} g^2\epsilon_{ijk}\epsilon_{abc}W^a_iW^b_jW^c_k &= [(M_{11})_{kc} + (M_{22})_{kc} + 2(M_{23})_{kc} + 2(M_{13})_{kc}] \\ &\quad\times [\mathcal{A}_{kc} + \mathcal{B}_{kc} + \mathcal{C}_{kc}] \\ &= [\mu_{kc} + \nu_{kc} + \tau_{kc}][\mathcal{A}_{kc} + \mathcal{B}_{kc} + \mathcal{C}_{kc}], \end{aligned} \quad (3.98)$$

where we have again used the abbreviation

$$\mu_{kc} \equiv \{(M_{11})_{kc} + (M_{22})_{kc}\}, \quad \nu_{kc} \equiv 2(M_{23})_{kc}, \quad \tau_{kc} \equiv 2(M_{13})_{kc}. \quad (3.99)$$

We then have

$$\begin{pmatrix} \mu_{kc}\mathcal{A}_{kc} & \mu_{kc}\mathcal{B}_{kc} & \mu_{kc}\mathcal{C}_{kc} \\ \nu_{kc}\mathcal{A}_{kc} & \nu_{kc}\mathcal{C}_{kc} & \nu_{kc}\mathcal{C}_{kc} \\ \tau_{kc}\mathcal{A}_{kc} & \tau_{kc}\mathcal{B}_{kc} & \tau_{kc}\mathcal{C}_{kc} \end{pmatrix} = \begin{pmatrix} 0 & 0 & 2\Theta'r^2[A^2 + r^2B^2] \\ 0 & 4B^2r^4\Theta' & 0 \\ 4\Theta'A^2r^2 & 0 & 0 \end{pmatrix}, \quad (3.100)$$

so we can therefore write

$$\frac{g^2r^2}{24\pi}\epsilon_{ijk}\epsilon_{abc}W^a_iW^b_jW^c_k = \frac{1}{4\pi}\Theta'r^4A^2 + \frac{1}{4\pi}r^6B^2\Theta'. \quad (3.101)$$

In total we have

$$\begin{aligned} Q = -\frac{1}{4\pi}\int_{t=t_0} \mathrm{d}r[&2BA'r^5 + 4ABr^4 - 2r^4A[B'r + 3B] \\ &+ 4A\Theta'r^2 + 2\Theta'r^4A^2 + 2r^6B^2\Theta']. \end{aligned} \quad (3.102)$$

Simplifying the above equation we can then can the following solvable integral for the Chern–Simons number

$$\begin{aligned} Q(\text{sphaleron}) &= \frac{1}{2\pi}\int_{\Theta(\infty)}^{\Theta(0)} d\Theta[\Theta - \cos\Theta] \\ &= \frac{1}{2} \end{aligned} \tag{3.103}$$

as required.

So let us summarize. The sphaleron is an unstable quasi-particle due to localization in time and it must continuously go between vacuum states. It is a finite-energy solution so it is well-behaved at the spatial boundary. The simplest solution is then derived by mapping the external space onto the internal space and coupling to a radial function that gives the desired properties and is determined by insisting that the sphaleron be a solution to the classical equations of motion. Such a sphaleron would be a local extremum of the action for any gauge group with an SU(2) subgroup. It has the nice property that the sphaleron rate is unsuppressed when electroweak symmetry is restored. It also has the unusual property that the Chern–Simons number is fractional and shown to be to equal 1/2. That it is non-zero means it facilitates baryon number non-conservation.

References

[1] Coon S A and Holstein B R 2002 Anomalies in quantum mechanics: the $1/r^2$ potential *Am. J. Phys.* **70** 513

[2] Holstein B R 2014 Understanding an anomaly *Am. J. Phys.* **82** 591

[3] Adler S 1969 Axial-vector vertex in spinor electrodynamics *Phys. Rev.* **177** 2426

[4] Bell J S and Jackiw R 1969 A PCAC puzzle: π 0 → γ γ in the omega-model *Nuovo Cimento* A **60** 47

[5] Jegerlehner F 2001 Facts of life with γ_5 *Eur. Phys. J.* C **18** 673–9

[6] El-Menoufi B K and White G A 2015 The axial anomaly, dimensional regularization and Lorentz-violating QED arXiv:1505.01754(hep-th)

[7] Ferrari R 2014 Managing γ_5 in dimensional regularization and ABJ anomaly arXiv:1403.4212

[8] Fujikawa K 1979 Path-integral measure for gauge-invariant fermion theories *Phys. Rev. Lett.* **42** 1195

[9] Donoghue J F, Golowich E and Holstein B R 2014 *Dynamics of the Standard Model* vol 35 (Cambridge: Cambridge University Press)

[10] Rajaraman R 1982 *Solitons and Instantons: An Introduction to Solitons and Instantons in Quantum Field Theory* (Amsterdam: North-Holland)

[11] Coleman S 1988 *Aspects of Symmetry: Selected Erice Lectures* (Cambridge: Cambridge University Press)

[12] Manton N and Sutcliffe P 2004 *Topological Solitons* (Cambridge: Cambridge University Press)

[13] Chern S and Simons J 1974 Characteristic forms and geometric invariants *Ann. Math.* **99** 48–69

[14] 't Hooft G 1976 Symmetry breaking through Bell–Jackiw anomalies *Phys. Rev. Lett.* **37** 8

[15] Wilczek F 1976 Inequivalent embeddings of SU(2) and instanton interactions *Phys. Lett.* B **65** 160
[16] Belavin A A, Polyakov A M, Schwartz A S and Tyupkin Y S 1975 Pseudoparticle solutions of the Yang–Mills equations *Phys. Lett.* B **59** 79
[17] Jackiw R and Rebbi C 1976 Vacuum periodicity in a Yang–Mills quantum theory *Phys. Rev. Lett.* **37** 172
[18] Callan C G, Dashen R F and Gross D J 1976 The structure of the gauge theory vacuum *Phys. Lett.* B **63** 334
[19] Polyakov A M 1977 Quark confinement and topology of gauge theories *Nucl. Phys.* B **120** 429
[20] Manton N S 1983 Topology in the Weinberg–Salam theory *Phys. Rev.* D **28** 8
[21] Klinkhamer F R and Manton N S 1984 A saddle-point solution in the Weinberg–Salam theory *Phys. Rev.* D **30** 10
[22] Cline J 2006 Baryogenesis arXiv:hep-ph/0609145

IOP Publishing

Electroweak Baryogenesis (Second Edition)
An introduction
Graham White

Chapter 4

Phase transitions

When the early Universe was at a very high temperature, the ground state of the Higgs potential changed [1]. For temperatures well above the weak scale the ground state is $\langle h \rangle = 0$ and electroweak symmetry is restored. As the Universe cools, a phase transition occurs when it becomes energetically favorable for electroweak symmetry to be spontaneously broken. Since sphalerons, a baryon number violating process, are unsuppressed in the high temperature phase, this phase transition becomes of central focus when studying electroweak baryogenesis [2–4].

This phase transition fulfils Sakharov's condition that there must be a departure from equilibrium. However, it turns out that it is insufficient to have an electroweak phase transition. The way the phase transition occurs can qualitatively affect the amount of baryon asymmetry. If the phase transition is second-order or cross-over, the sphalerons which you rely on for baryon number violation work against you by washing out any asymmetry. What is needed to produce a large enough baryon asymmetry is a strongly first-order phase transition which is characterized by a medium of symmetric vacuum with bubbles where electroweak symmetry is broken. Such bubble nucleation occurs when there is a barrier between the new and old minima in the effective potential so that the transition needs to occur by tunneling, as depicted in figure 4.1.

If there is a strong source of CP violation then interactions with the bubble wall produce the baryon asymmetry, which is then frozen in as the bubble grows to become the observable Universe. If the suppression of sphalerons within the bubble is strong enough, the baryon asymmetry formed in front of its advancing wall is swept up into the new phase and frozen in. The order of the electroweak phase transition within the Standard Model is dependent on the mass of the Higgs where the order of the phase transition becomes second-order and eventually cross-over for a heavier Higgs. For a Higgs mass of 125 GeV the electroweak phase transition is cross-over. The order of the phase transition can be made first-order by additional scalar fields. However, a first-order phase transition is necessary but not sufficient

doi:10.1088/978-0-7503-3571-3ch4

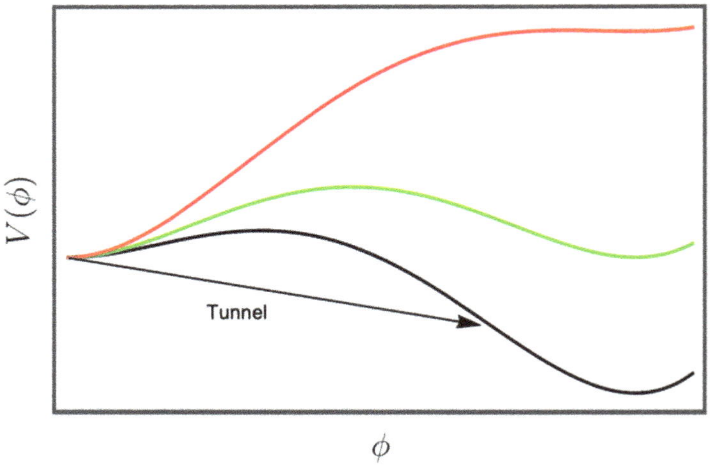

Figure 4.1. Schematic of strongly first-order phase transition. Such a phase transition is characterized by a barrier separating two phases and the barrier persisting when the new phase is lower than the old. In such a case the transition proceeds via tunneling processes. There are three temperatures of interest, the critical temperature where the minima are degenerate (green line), the temperature some time above the critical temperature where the minimum first appears (red line), and the temperature somewhere below the critical temperature where tunneling rates are fast enough that there is at least one critical bubble per Hubble volume (the nucleation temperature).

for sphaleron suppression inside the bubble. We need the sphaleron rate to be very small within the bubble and since the sphaleron rate is suppressed by a factor of $\exp[-v_H(T)/T \times \cdots]$ we can parametrize the strength of the phase transition by $v_H(T_C)/T_C$, where T_C is the critical temperature, defined as the temperature at which the vacuum has a degenerate minimum between the origin and the electroweak symmetry breaking phase. Numerically, one can derive an approximate condition for baryon particle number conservation of [5]

$$v_H(T_C)/T_C \gtrsim 1. \tag{4.1}$$

Some time below the critical temperature is the temperature at which a bubble forms that is of a critical size which grows to become the observable Universe (colliding with other critical bubbles in the process). This temperature, known as the nucleation temperature, will be defined as T_N. The crucial quantities we will need to calculate are therefore:

- the critical temperature;
- the order and strength of the phase transition;
- the nucleation temperature;
- the space–time bubble wall profile;
- the nucleation rate; and
- the sphaleron rate.

The best estimates of each of these is performed using non-perturbative Monte Carlo methods. However, such methods are so difficult to apply that only a few specific models of electroweak symmetry breaking have been studied using such methods. We therefore focus on finite temperature perturbative techniques. Such techniques provide a reasonable approximation of the more rigorous non-perturbative methods giving qualitative information about the relationship of the above list of quantities on model parameters. However, any calculation involving perturbative finite temperature methods should be assumed to contain a large amount of theoretical uncertainty. The temperature at which the phase transition occurs within the Standard Model, for example, is significantly lower when calculated non-perturbatively compared to the value obtained through perturbative techniques. We touch on some of these uncertainties in this chapter and go into more detail in chapter 9.

The pre-requisite knowledge for performing these calculations is finite temperature QFT and we can progress no further without a review. Finite temperature QFT is a very broad subject so we will limit our discussion to only the key concepts necessary for calculations in this field while keeping things as self-contained as possible. We will assume knowledge of the effective potential and its derivation at zero temperature. After giving a small pedagogical review of finite temperature field theory we will derive the effective potential at finite temperature and calculate the order of the phase transition and its order in the Standard Model. We will then demonstrate the need for the baryon particle number conservation condition to be met before explaining methods of making it strongly first-order through the introduction of new scalar fields. Some issues of gauge invariance will be discussed. Then we will give an overview on how the vacuum decays from a false vacuum to true vacuum, how to calculate the tunneling rate of this process, and the nucleation temperature.

4.1 Closed time path formalism

To describe the particle dynamics during the electroweak phase transition we need to reformulate QFT slightly to handle the out-of-equilibrium nature of this process. At equilibrium we define states at the infinite past and use the usual machinery of QFT to then predict the 'out states' at the infinite future. Out of equilibrium we do not know the infinite future. What we can do is analyze a system which has been in equilibrium until a point $t = 0$ where the system departs from equilibrium. The in-states can then be evolved to the point where the system departs from equilibrium and we can evolve the system from there [6]. Recall that at finite temperate the density matrix of a system is described as

$$\rho(0) = \frac{e^{-\beta\mathcal{H}}}{\mathrm{Tr}e^{-\beta\mathcal{H}}}, \tag{4.2}$$

where the argument of the density matrix is given to indicate it is at some time $t = 0$. When our system departs from equilibrium, our density matrix acquires a time dependence in the usual way

$$\rho(t) = U(t, 0)\rho(0)U^{\dagger}(t, 0) \tag{4.3}$$

with the time evolution unitary operator defined

$$
\begin{aligned}
i\frac{\partial U}{\partial t} &= H(t)U(t, t') \\
U(t, t') &= T\left(\mathrm{e}^{-i\int_{t'}^{t} \mathrm{d}t'' H(t'')}\right).
\end{aligned}
\tag{4.4}
$$

Note that from the above we obtain the relation that $U(t_1, t_2)^\dagger = U(t_2, t_1)$ and $U(t_1, t_2)U(t_2, t_3) = U(t_1, t_3)$ when $t_1 > t_2 > t_3$. Let the system be in equilibrium up to $t = 0$ meaning that there is no time dependence in the Hamiltonian up to a time where the system goes out of equilibrium (say at the electroweak phase transition) at the time $t = 0$. Note that

$$
T\left(\mathrm{e}^{-i\int_{T-i\beta}^{T} \mathrm{d}t\mathcal{H}(t)}\right) = T\left(\mathrm{e}^{-i\int_{T-i\beta}^{T} \mathrm{d}t\mathcal{H}_0}\right) = \mathrm{e}^{-\beta\mathcal{H}}. \tag{4.5}
$$

We can therefore write the equilibrium and non-equilibrium density matrices as

$$
\begin{aligned}
\rho(0) &= \frac{U(T - i\beta)}{\mathrm{Tr}\ U(T - i\beta, T)} \\
\rho(t) &= \frac{U(t, 0)U(T - i\beta)U(0, t)}{\mathrm{Tr}\ U(T - i\beta, T)}.
\end{aligned}
\tag{4.6}
$$

The time-dependent expectation value of an operator is defined in the usual way $\langle A(t)\rangle = \mathrm{Tr}\,\rho(t)A$. Let $T < 0$ and $T' \gg 0$. Unitary evolution operators with negative values in both time arguments commute with each other since $\mathcal{H}_0$ commutes with itself. We can write the time-dependent expectation value of an operator as

$$
\begin{aligned}
\langle A(t)\rangle &= \frac{Tr\ [U(t, 0)U(T - i\beta)U(0, t)A_0]}{\mathrm{Tr} U(T - i\beta, T)} \\
&= \frac{Tr[U(t, 0)U(0, T)U(T, 0)U(T - i\beta)U(0, t)A_0]}{\mathrm{Tr}\ U(T - i\beta, T)} \\
&= \frac{Tr\ [U(t, T)U(T - i\beta)U(T, t)A_0]}{\mathrm{Tr}\ U(T - i\beta, T)} \\
&= \frac{Tr\ [U(t, T)U(T - i\beta)U(T, T')U(T', T)U(T, t)A_0]}{\mathrm{Tr}\ U(T - i\beta, T)U(T, T')U(T', T)} \\
&= \frac{Tr\ [U(T - i\beta, T)U(T, T')U(T', t)AU(t, T)]}{\mathrm{Tr}\ U(T - i\beta, T)U(T, T')U(T', T)}.
\end{aligned}
\tag{4.7}
$$

Let us take the limit as $T \to \infty$. Relating the above equation to the type of calculations we wish to perform in QFT (e.g. expectation values of correlation functions), we are performing a different time contour to the standard equilibrium case. Our time contour runs from the infinite past to some time at the present T' where the system is out of equilibrium back into the infinite past before following a path along imaginary time. This is shown in figure 4.2 and is known as the closed time path contour [7–13]. Recall that a generating functional in QFT is given by a source-dependent partition function, say

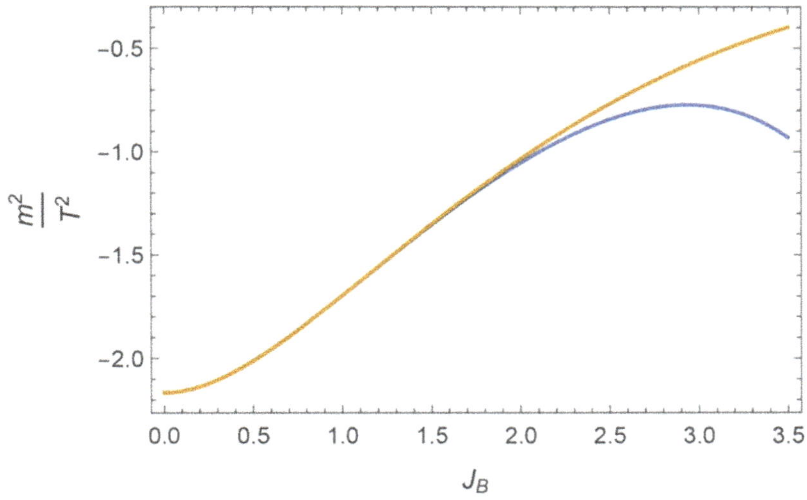

Figure 4.2. $J_B(m^2/T^2)$ compared to its high temperature expansion. The approximation works for $m^2/T^2 \lesssim 2$. The golden line denotes the full thermal function, whereas the blue line denotes the high temperature expansion.

$$\begin{aligned} Z(\beta) &= \mathrm{Tr} e^{-\beta \mathcal{H}} \\ Z(\beta, \mathcal{J}) &= N \int \mathcal{D}\phi e^{-S_E - \mathcal{J}\phi}. \end{aligned} \tag{4.8}$$

We can then identify the denominator in the last line of equation (4.7) with the generating functional in our out of equilibrium description by suitably adjusting the unitary to include a source term defined along the contour C_i given in figure 4.3

$$\begin{aligned} Z[\beta, \mathcal{J}_c] &= \mathrm{Tr}\ U_{\mathcal{J}_c}(T - i\beta, T) U_{\mathcal{J}_c}(T, T') U_{\mathcal{J}_c}(T', T) \\ &= \int \mathcal{D}\phi e^{i\int_c \mathrm{d}t \int \mathrm{d}^3x(\mathcal{L} + \mathcal{J}_c\phi)}. \end{aligned} \tag{4.9}$$

Examining the closed time contour, if we wish to form bilinear correlation functions we have four possibilities depending on whether both field operators are on the positive imaginary side of the contour, both are on the negative side, or the two cases where you have one on each. For a scalar field these four cases are

$$\begin{aligned} i\Delta^{>}(x, y) &= \langle \phi(x)\phi^{\dagger}(y) \rangle \\ i\Delta^{<}(x, y) &= \langle \phi^{\dagger}(x)\phi(x) \rangle \\ i\Delta^{T}(x, y) &= \langle T[\phi(x)\phi^{\dagger}(y)] \\ &= \Theta(x_0 - y_0) i\Delta^{>}(x, y) + \Theta(y_0 - x_0) i\Delta^{<}(x, y) \\ i\Delta^{\bar{T}}(x, y) &= \langle \bar{T}[\phi(x)\phi^{\dagger}(y)] \\ &= \Theta(x_0 - y_0) i\Delta^{<}(x, y) + \Theta(y_0 - x_0) i\Delta^{>}(x, y). \end{aligned} \tag{4.10}$$

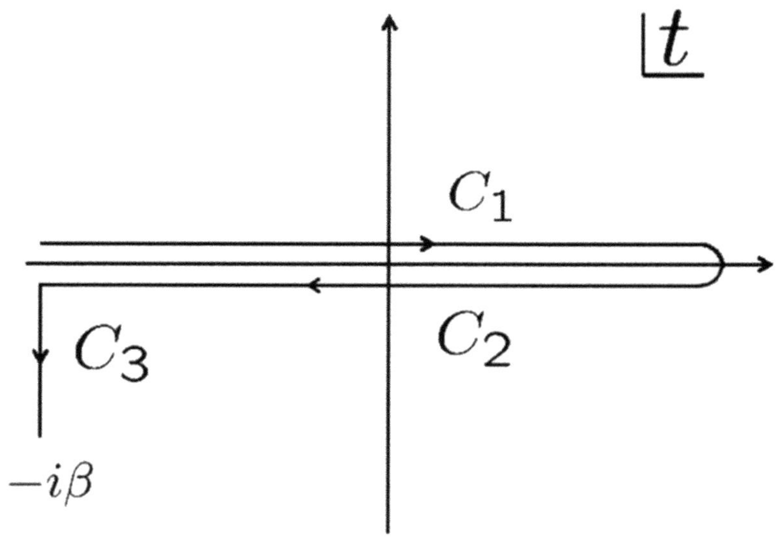

Figure 4.3. The closed time path contour. The contour runs parallel and infinitesimally above the real line from minus infinity to the present, runs parallel and infinitesimally below the real line back to minus infinity, before running parallel to the imaginary axis to $-i\beta$.

To derive the explicit form of these propagators, we first note that $\mathcal{H} \sim a^{\dagger}a$ and that therefore $\mathrm{Tr}[\mathrm{e}^{\beta\mathcal{H}}] = (1 - \mathrm{e}^{\beta\omega})^{-1} = n_B$. This means that $a^{\dagger}a = n_B(\omega_p)\delta^3(p - k)$. Also, it will be more convenient to work in momentum space. In the case of spatial isotropy and homogeneity, the momentum space representation version of the propagators is a function of a single momentum variable. The most general form for our propagators can be written as the sum of the zero temperature propagators plus some as yet unspecified momentum conserving, temperature-dependent corrections

$$\begin{pmatrix} \Delta^T(p) & \Delta^<(p) \\ \Delta^<(p) & \Delta^{\bar{T}}(p) \end{pmatrix} = \begin{pmatrix} \dfrac{1}{p^2 - M^2 + i\epsilon} & 0 \\ 0 & \dfrac{-1}{p^2 - M^2 - i\epsilon} \end{pmatrix} + \begin{pmatrix} \tilde{c}_1(p)\delta(p^2 - M^2) & \tilde{c}_{1_3}(p)\delta(p^2 - M^2) \\ \tilde{c}_{1_2}(p)\delta(p^2 - M^2) & \tilde{c}_{1_4}(p)\delta(p^2 - M^2) \end{pmatrix}. \tag{4.11}$$

We will then argue the full form of these propagators using the constraints of unitarity, Hermiticity and CPT invariance [14][1]. Let us begin with unitarity. It is straightforward to show that either using unitarity or the definitions of the scalar propagators given in equation (4.10) the following identity holds

$$\begin{aligned} i\Delta^T(x, y) + i\Delta^{\bar{T}}(x, y) &= i\Delta^<(x, y) + i\Delta^>(x, y) \\ i\Delta^T(p) + i\Delta^{\bar{T}}(p) &= i\Delta^<(p) + i\Delta^>(p). \end{aligned} \tag{4.12}$$

[1] Causality is also indirectly implied by the form of our propagators.

This, in combination with the identity

$$\frac{1}{p^2 - M^2 + i\epsilon} - \frac{1}{p^2 - M^2 - i\epsilon} = -2\pi i\delta(p^2 - M^2), \tag{4.13}$$

gives the relation

$$-2\pi i + (\tilde{c}_1 + \tilde{c}_4) = \tilde{c}_2 + \tilde{c}_3. \tag{4.14}$$

Then we can use CPT invariance and Hermiticity to further constrain these functions. To use CPT invariance we recall the effect of charge conjugation operator acting on a scalar field

$$\begin{aligned} U_C^\dagger \phi(x) U_C &= e^{i\eta}\phi^\dagger(x) \\ U_C^\dagger \phi^\dagger(x) U_C &= e^{-i\eta}\phi(x). \end{aligned} \tag{4.15}$$

We then obtain the following relationships between the coefficients in our generic form of our propagator

$$\begin{aligned} \Delta^>(p) &= -\Delta^{*>}(p) = \Delta^{C<}(-p) = -\Delta^{*C<}(-p) \\ \Delta^T(p) &= \Delta^{CT}(-p) = -\Delta^{C\bar{T}}(p) = -\Delta^{*C\bar{T}}. \end{aligned} \tag{4.16}$$

These relations give the relations

$$\begin{aligned} \tilde{c}_1(p) &= \tilde{c}_1(-p),\ \tilde{c}_4(p) = \tilde{c}_4(-p),\ \tilde{c}_2(p) = \tilde{c}_3(-p) \\ \tilde{c}_1(p) &= -\tilde{c}_4^*(p),\ \mathrm{Re}[\tilde{c}_2(p)] = \mathrm{Re}[\tilde{c}_3(p)] = 0. \end{aligned} \tag{4.17}$$

The most general solution to the above solutions consistent with isotropy and spatial homogeneity[2] and taking note of the fact that $a^\dagger a = n_B(|\,p_0\,|)\delta^3(p - k)$ is

$$\begin{aligned} &\begin{pmatrix} \tilde{c}_1(p)\delta(p^2 - M^2) & \tilde{c}_3(p)\delta(p^2 - M^2) \\ \tilde{c}_2(p)\delta(p^2 - M^2) & \tilde{c}_4(p)\delta(p^2 - M^2) \end{pmatrix} \\ &= -2\pi i\left[n_B(|\,p_0\,|)\begin{pmatrix} 1 & 1 \\ 1 & 1 \end{pmatrix} + \begin{pmatrix} 0 & \Theta(-p_0) \\ \Theta(p_0) & 0 \end{pmatrix}\right]. \end{aligned} \tag{4.18}$$

A similar derivation for fermions gives that the correction to the zero temperature propagators are given by

$$\begin{aligned} \begin{pmatrix} S^T(p) & S^<(p) \\ S^>(p) & S^{\bar{T}} \end{pmatrix} &= \begin{pmatrix} \dfrac{\not{p} + m}{p^2 - m^2 + i\epsilon} & 0 \\ 0 & \dfrac{\not{p} + m}{p^2 - m^2 - i\epsilon} \end{pmatrix} \\ &+ 2\pi i(\not{p} + m)\left[n_F(|\,p_0\,|)\begin{pmatrix} 1 & 1 \\ 1 & 1 \end{pmatrix} + \begin{pmatrix} 0 & -\Theta(-p_0) \\ -\Theta(p_0) & 0 \end{pmatrix}\right]. \end{aligned} \tag{4.19}$$

[2] We also imply causality though we have not explicitly discussed it here.

The propagators and the closed time path contour turn out to be all we need to understand to manipulate the effective potential at high temperature.

4.2 A brief review of the effective potential at zero temperature

The effective potential receives finite temperature corrections at loop level that can qualitatively change its minima and therefore the vacuum expectation value. This topic at zero temperature is generally treated in graduate level courses [5, 15, 27], however, diving straight into the finite temperature treatment of the effective potential can be disorientating. Therefore, to warm up we will review how to calculate the one-loop corrections to a simple scalar field theory at zero temperature. We also omit contributions from fermions based on the fact that this topic is elementary and just including the scalar self-interactions is a compromise between not covering too much elementary review and easing the transition to finite temperature calculations. The goal of this short review is threefold—to demonstrate that the one-loop correction to the effective potential can be written as a geometric series in the propagator, to review what the forms of the one-loop corrections to the effective potential look like after regularization and renormalization, and to introduce a trick that simplifies the calculation by taking the derivative of the one-loop correction to the effective potential and integrating. To derive the effective potential we begin as usual from the generating functional

$$iW_i[j] \equiv \log z[j] = \log\left[\int \mathrm{d}\phi \, \exp(S[\phi] + \phi j \exp)\right]. \tag{4.20}$$

Here spatial integrals and arguments have been suppressed for notational convenience. As indicated by the argument, the natural variable of $W[j]$ is j. That is,

$$\mathrm{d}W = \frac{\partial W}{\partial j}\mathrm{d}j = \phi \mathrm{d}j. \tag{4.21}$$

To change the natural variables of W from j to ϕ, one simply performs a Legendre transformation:

$$\Gamma[\phi] = W[j] - \phi j. \tag{4.22}$$

The left-hand side of the above equation is the effective action. The effective potential for a translational invariant theory is defined by calculating Γ for a constant field configuration

$$\Gamma(\phi_c) = -\int \mathrm{d}^4 x \, V_{\text{eff}}(\phi_c). \tag{4.23}$$

To calculate the effective potential, we can Taylor expand the effective action in powers of the field

$$\Gamma[\phi] = \sum_{n=0}^{\infty} \frac{\phi^n}{n!} \frac{\partial^n \Gamma(\phi)}{\partial \phi}\bigg|_{\phi=0}. \tag{4.24}$$

Let us now restore the spatial dependencies and integrals

$$\Gamma[\phi] = \sum_{n=0}^{\infty} \frac{1}{n!} \int \prod_{i=1}^{n} \mathrm{d}^4 x_i \phi(x_i) \Gamma^n(x_1 \cdots x_n). \tag{4.25}$$

Here $\Gamma^n(x)$ are the one-particle irreducible Green's functions. For a scalar field theory with quartic interactions these functions are given in figure 4.4. Let us take the Fourier transform of these Green's functions.

$$\Gamma^n(x) = \int \left(\prod_{i=1}^{n} \frac{\mathrm{d}^4 p_i}{(2\pi)^4} \exp(ip_i x_i) \right) (2\pi)^4 \delta^4(p_1 \cdots + p_n) \Gamma^n(p), \tag{4.26}$$

where

$$\phi(p) = \int \mathrm{d}^x \mathrm{e}^{-ipx} \phi(x). \tag{4.27}$$

So we have as our expression for the effective action

$$\Gamma(\phi_c) = \frac{1}{n!} \sum_{n=0}^{\infty} \int \prod_{i=1}^{n} \left[\frac{\mathrm{d}^4 p_i \phi(-p_i)}{(2\pi)^4} \right] (2\pi)^4 \delta^{(4)}(p_1 + \cdots + p_n) \Gamma^n(p_1 + \cdots + p_n)_{\phi_c}. \tag{4.28}$$

For a simple example that gives us most of the essential physics consider a model with a single scalar field whose tree-level potential is given by

$$V_0 = \frac{1}{2} m^2 \phi^2 + \frac{\lambda}{4!} \phi^4. \tag{4.29}$$

Our goal is to calculate the effective potential to one loop. To do so we will set the scalar field to be a constant in space–time as stated before. In momentum space the scalar field is therefore

$$\phi_c(p) = (2\pi)^4 \phi_c \delta^4(p). \tag{4.30}$$

Substituting this into the expression we have for the effective potential gives

$$\Gamma[\phi_c] = -\sum_{n=0}^{\infty} \frac{\phi_c^n}{n!} \Gamma^n(p_i = 0). \tag{4.31}$$

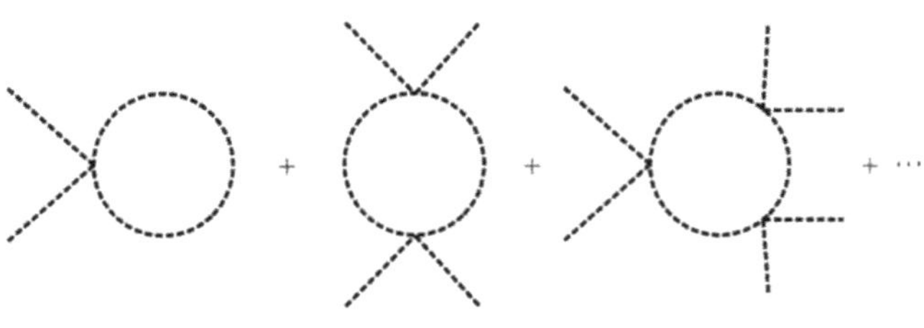

Figure 4.4. One-loop contributions to the effective potential in free scalar field theory. Reproduced from [5].

Using the explicit form for one-particle irreducible diagrams that are at most one loop from figure 4.4 we see that the above sum is actually a geometric series

$$\begin{aligned} V_1 &= i\sum_{n=0}^{\infty}\int \frac{d^4p}{(2\pi)^4}\frac{1}{2n}\left[\frac{\lambda\phi_c^2/2}{p^2 - m^2 + i\epsilon}\right]^n \\ &= \frac{1}{2}\int \frac{d^4p}{(2\pi)^4}\log\left[p^2 + m^2(\phi)\right]. \end{aligned} \tag{4.32}$$

We can calculate this easily using a hard cutoff regulator

$$\begin{aligned} V_1 &= \frac{1}{32\pi^2}\int_0^{\Lambda^2} p^2 \log\left[p^2 + m^2\right]dp^2 \\ &= \frac{1}{32\pi^2}\left[\frac{1}{2}(p^4 - m^4)\log(p^2 + m^2) - \frac{1}{2}\left(\frac{p^4}{2} - p^2m^2\right)\right]_0^{\Lambda^2}. \end{aligned} \tag{4.33}$$

So in total we have the one-loop correction to the effective potential at zero temperature

$$V_1 = \frac{1}{32\pi^2}\left[\frac{1}{2}(\Lambda^4 - m^4)\log(\Lambda^2 + m^2) - \frac{1}{2}\left(\frac{\Lambda^4}{2} - \Lambda^2m^2\right) + \frac{m^4}{2}\log(m^2)\right]. \tag{4.34}$$

We can expand the first log term into two logarithms $\log(\Lambda^2 + m^2) = \log(\Lambda^2) + \log(1 + m^2/\Lambda^2)$. The second logarithm can be expanded to second-order. Taking the limit $\Lambda\to\infty$ and ignoring m-independent terms we can write

$$V = \frac{1}{32\pi^2}m^2\Lambda^2 + \frac{1}{64\pi^2}m^4\left[\log\frac{m^2}{\Lambda^2} - \frac{1}{2}\right]. \tag{4.35}$$

Finally, we want to revise how to renormalize the one-loop zero temperature effective action using two schemes of regularization—finite cutoff regularization as used above and dimensional regularization. Let us introduce counter terms into our Lagrangian

$$V = \frac{1}{2}\delta m^2\phi_c^2 + \frac{\lambda + \delta\lambda}{4!}\phi_c^4 + \frac{\lambda\phi_c^2\Lambda^2}{64\pi^2} + \frac{\lambda^2\phi_c^4}{256\pi^2}\left(\log\frac{\lambda\phi_c^2}{\Lambda^2} - \frac{1}{2}\right). \tag{4.36}$$

We are free to choose a renormalization scheme where

$$\left.\frac{d^2V}{d\phi^2}\right|_{\phi=0} = 0, \qquad \left.\frac{d^4V}{d\phi^4}\right|_{\phi=\mu} = \lambda \tag{4.37}$$

which gives the conditions

$$\delta m^2 = \frac{\lambda\Lambda^2}{64\pi^2} \tag{4.38}$$

$$\delta\lambda = -\frac{\lambda^2}{32\pi^2} - \frac{3\lambda^2}{32\pi^2}\log\frac{\lambda\mu^2}{2\Lambda^2}. \tag{4.39}$$

From this we obtain the renormalized effective potential using a cutoff regularization scheme

$$V_{\text{eff}} = \frac{\lambda\phi^4}{4!} + \frac{\lambda^2\phi^4}{256\pi^2}\left(\log\frac{\phi^2}{\mu^2} - \frac{25}{6}\right). \tag{4.40}$$

We will conclude by calculating the renormalized effective potential in dimensional regularization and showing a trick which simplifies the calculation considerably. Our one-loop correction to our potential for $d = 4 - \epsilon$ is

$$V_1(\phi_c) = \frac{1}{2}(\mu^2)^{2-\frac{n}{2}}\int\frac{\mathrm{d}^n p}{(2\pi)^n}\log[p^2 + m^2]. \tag{4.41}$$

It is much easier to calculate the one-loop correction to the effective potential by first calculating it with respect to the mass and then integrating. The derivative is just a single disconnected bubble

$$\begin{aligned}\frac{\partial V_1}{\partial m^2} &= \frac{1}{2}(\mu^2)^{2-\frac{n}{2}}\int\frac{\mathrm{d}^n p}{(2\pi)^n}\frac{1}{p^2 + m^2}\\ V_1 &= \frac{m^4}{64\pi}\left(-\left[\frac{1}{\epsilon} - \gamma_E + \log 4\pi\right] + \log\frac{m^2}{\mu^2} - \frac{3}{2}\right),\end{aligned} \tag{4.42}$$

where we have used the relation $A^\epsilon \approx 1 - \epsilon\log A$. We can subtract the $1/\epsilon - \gamma_E - \log 4\pi$ term to obtain

$$V_1 = \frac{1}{64\pi^2}m^4\left(\log\frac{m^2}{\mu^2} - \frac{3}{2}\right), \tag{4.43}$$

where m^2 is the tree-level field-dependent mass defined by $m^2 = \partial^2 V_0(\phi)$. Here we have used a mass-independent renormalization scale μ. The perturbative expansion is most accurate when μ is about the scale of the maximum value of $m_i(\phi)$.

4.3 The effective potential at finite temperature

From the above revision of the zero temperature effective potential one can guess that the derivation of the one-loop effective potential at finite temperature involves summing over all one-particle irreducible interactions at finite temperature. This can be quite cumbersome—particularly in the closed time path formalism—as there are now four Green's functions instead of one. However, a dramatic simplification can be made by using the trick at the end of the previous subsection where we observed that the derivative of one-loop corrections to shifted effective potentials, $\partial V/\partial m^2(\phi_c)$, are diagrammaticaly equivalent to disconnected bubble diagrams. One can then calculate this diagram(s) and integrate. Returning to the simple example of ϕ^4 theory, we have only one bubble diagram to compute and in fact only one propagator to concern ourselves with, since external legs must be on the positive time contour (that is the contour that goes from $-\infty$ to the present ϵ above the real axis):

$$\frac{\partial V}{\partial m^2(\phi_c)} = \frac{1}{2}\int \frac{\mathrm{d}^4p}{(2\pi)^4}\left[\frac{i}{p^2 - m^2(\phi) + i\epsilon} + 2\pi n_B(\omega_p)\delta(p^2 - m^2)\right]. \tag{4.44}$$

So we see the one-loop correction is a sum of the zero temperature component and the finite temperature component. Let us focus exclusively on the finite temperature component. The time integral can be calculated by using the identity

$$\delta(p^2 - m^2) = \frac{1}{2\omega_p}\left[\delta(p_0 + \omega_o) + \delta(p_0 - \omega_p)\right], \tag{4.45}$$

which allows us to write the finite temperature part of the bubble

$$\frac{\partial V_1(T)}{\partial m^2(\phi_c)} = \int \frac{\mathrm{d}^3p}{(2\pi)^3}\frac{1}{2\omega}n_B(\omega). \tag{4.46}$$

Let us make the convenient substitution $p = Tx$. It is easy to show by direct substitution that the integral of the above equation is [16]

$$V_1(T) = \frac{T^4}{2\pi^2}J_B\left(\frac{m^2(\phi_x)}{T^2}\right) \equiv \frac{T^4}{2\pi^2}\int_0^\infty \mathrm{d}x x^2 \log\left[1 - \mathrm{e}^{-\sqrt{x^2+\frac{m^2(\phi_c)}{T^2}}}\right]. \tag{4.47}$$

This is just the free energy of a relativistic gas of bosons. Next we turn our attention to fermion contributions to the one-loop finite temperature effective potential. The process is identical to the above where one considers the derivative of the effective potential, $\partial V_1/\partial m_f(\phi_c)$, and integrates. In this case we obtain the expression

$$V_1(T) = \frac{T^4}{2\pi^2}J_F\left(\frac{m_f^2(\phi_x)}{T^2}\right) \equiv \frac{T^4}{2\pi^2}\int_0^\infty \mathrm{d}x x^2 \log\left[1 + \mathrm{e}^{-\sqrt{x^2+\frac{m_f^2(\phi_c)}{T^2}}}\right], \tag{4.48}$$

which is the free energy of a relativistic gas of fermions. In practice, these integral expressions can be cumbersome to work with. It is useful to make an expansion into a polynomial in the fields. There is a high temperature expansion for each case which is valid for relatively small m^2/T^2. For the bosonic thermal function the expansion goes as

$$J_B(z^2) \approx -\frac{\pi^4}{45} + \frac{\pi^2 z^2}{12} - \frac{\pi z^3}{6} - \frac{z^4}{32}\left(\log z^2 - 5.4076\right) + \Delta J_B(z^2). \tag{4.49}$$

As seen in figure 4.2, the high temperature expansion works quite well for arguments up to ~2. The higher order corrections

$$\Delta J_B(z^2) = -2\pi^{7/2}\sum_{l=1}^{\infty}(-1)^l\frac{\zeta(2l+1)}{(l+1)!}\Gamma\left(l + \frac{1}{2}\right)\left(\frac{z^2}{4\pi^2}\right)^{l+2} \tag{4.50}$$

extend the region of validity somewhat. This we show in figure 4.5, where the first two terms in this series moderately affect the region of validity. The high temperature fermion expansion

$$J_F(z^2) \approx \frac{7\pi^4}{360} - \frac{\pi^2 z^2}{24} - \frac{z^4}{32}\left(\log z^2 - 2.6351\right) + \Delta J_F(z^2) \tag{4.51}$$

similarly works fairly well for arguments up to about ~2, as seen in figure 4.6, but unfortunately the higher order terms

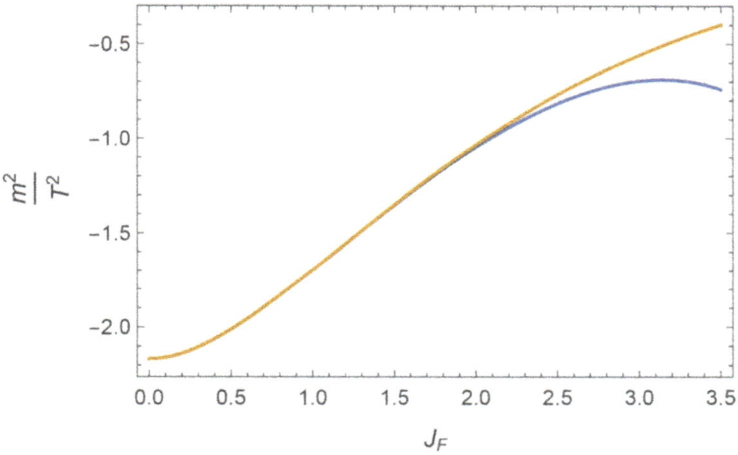

Figure 4.5. $J_B(m^2/T^2)$ compared to a high temperature expansion extended to terms of $O(m^8/T^8)$. The golden line again is the full thermal function, whereas the blue line is the high temperature expansion. The higher order terms make only a slight extension to the region of validity. Note the Riemann zeta function $\zeta(5)$ in equation (4.51) has been approximated to be 1. Fine-tuning this value makes little difference.

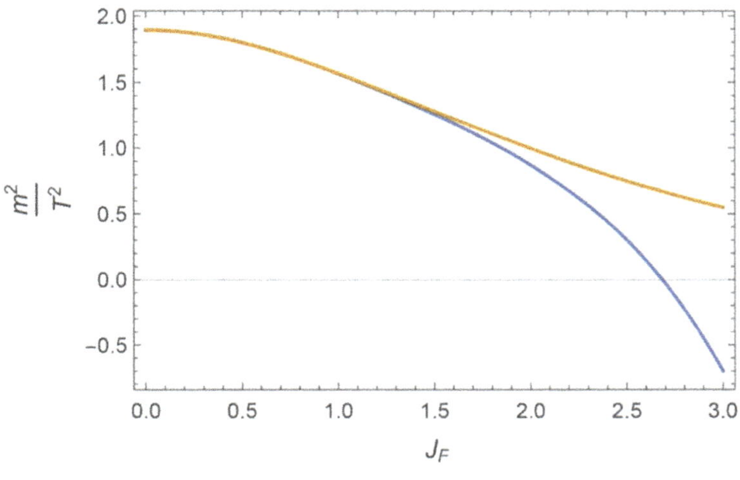

Figure 4.6. $J_F(m^2/T^2)$ compared to its high temperature expansion. The approximation works for $m^2/T^2 \lesssim 2$. The golden line is the full thermal function, whereas the blue line is the approximation.

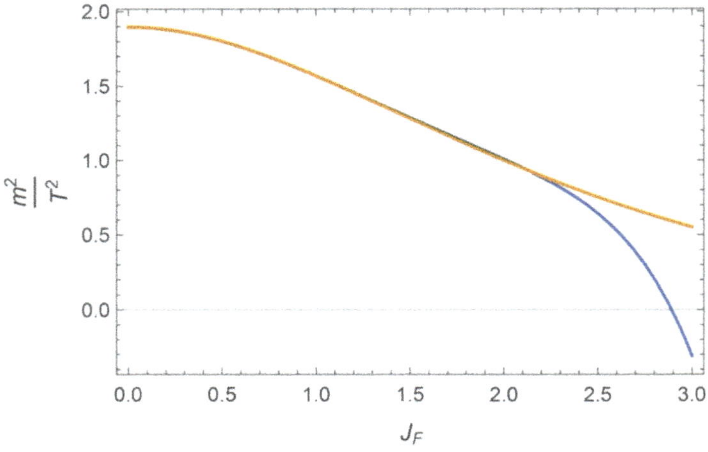

Figure 4.7. $J_F(m^2/T^2)$ compared to a high temperature expansion extended to terms of $O(m^8/T^8)$. The golden line is the full thermal function, whereas the blue line is the high temperature expansion. The higher order terms make only a slight extension to the region of validity. Note the Riemann zeta function $\zeta(5)$ in equation (4.53) has been approximated to be 1. Fine-tuning this value makes little difference.

$$\Delta J_F(z^2) = -\frac{\pi^{7/2}}{4}\sum_{l=1}^{\infty}(-1)^l \frac{\zeta(2l+1}{(l+1)!}(1 - 2^{-2l-1})\Gamma\left(l + \frac{1}{2}\right)\left(\frac{z^2}{\pi^2}\right)^{l+2} \tag{4.52}$$

suffer from the fact that they converge slowly, as shown in figure 4.7, where the first two correction terms barely extend the region of validity. When the field-dependent masses are very large compared to the temperature, the thermal functions contain Boltzmann suppression,

$$J_B(z^2) = J_F(z^2) \approx \left(\frac{z}{2\pi}\right)^{\frac{3}{2}} e^{-z}[1 + O(z^{-1})], \tag{4.53}$$

so we can safely ignore the contributions to the electroweak phase transition for very heavy particles.

4.3.1 The Standard Model example

The Standard Model is a simple enough regime that within it you can treat the electroweak phase transition analytically within the high temperature expansion, given a couple of approximations. The Standard Model Higgs is a scalar, complex, SU(2) doublet so it has four degrees of freedom which can be put into a column vector h_i. Let us first approximate the full finite temperature effective potential by only considering contributions from the massive gauge bosons, Goldstone bosons, and the top quark as these contributions dominate. We will work in the Feynman gauge, $\xi = 0$ for simplicity, returning to the issue of gauge invariance later. The mass expansion in this case is neat since these masses are all proportional to the vacuum expectation value. We also include the quadratic thermal correction to μ including contributions from Goldstone bosons. If we are loose with concerns of gauge invariance (an issue we return to in a later section) we can write

$$V(H, T) \approx \left(-\frac{\mu^2}{2} + c_h T^2\right)h^2 - ETh^3 + \frac{\lambda}{4}h^4 \tag{4.54}$$

for coefficients

$$\begin{aligned} c_h &= \frac{1}{32}\left(g_1^2 + 3g_2^2 + 4y_t^2 + 8\lambda\right) \\ E &= \frac{3}{96\pi}\left(2g_2^3 + \left(g_1^2 + g_2^2\right)3/2\right). \end{aligned} \tag{4.55}$$

Qualitatively one can describe from the form of the above equations how the Standard Model effective potential evolves with temperature. At some certain temperature $c_h T^2 > \mu^2$ and the coefficient of the bilinear term in the effective potential changes sign. The cubic term is negative definite, and if the parameters of the Standard Model occupy a particular range then the potential will have an additional turning point appear which is a local maximum between the origin and the global minimum. Outside such a range the minimum will evolve ever closer to the origin until it disappears. Since the bilinear term in the potential is quadratic in the temperature, whereas the cubic grows linearly with temperature, the third turning point will eventually disappear at very high temperature and the potential will have a single minimum at the origin.

Recall a first-order phase transition is characterized by a bump separating the origin and the global minimum, whereas a second-order phase transition is characterized by the new minimum just rolling away from the origin continuously as the system cools. At very high temperature the potential has a single minimum at the origin with no bumps or wiggles. At a lower temperature the minimum becomes degenerate. This is known as the critical temperature. At a much lower temperature the barrier disappears at which point the phase transition is well and truly complete. We can define the conditions for the critical temperature as

$$\begin{aligned} V(h_C, T_C) &= 0 \\ \left.\frac{\partial V(h, T)}{\partial h}\right|_{h_C, T_C} &= 0, \end{aligned} \tag{4.56}$$

which is easy to solve in the Standard Model. A simple calculation gives

$$\begin{aligned} h_C &= \frac{E}{\sqrt{c_h}}\frac{\mu^2}{\sqrt{\lambda(c_h\lambda - E^2)}} \\ T_C &= \frac{\lambda}{2\sqrt{c_h}}\frac{\mu^2}{\sqrt{\lambda(c_h\lambda - E^2)}}, \end{aligned} \tag{4.57}$$

which gives the simple relationship for the strength of the phase transition that can be written in terms of either the quartic coupling or the tree-level $0T$ Higgs mass

$$\frac{h_c}{T_C} = \frac{E}{2\lambda} = \frac{Ev_H^2}{m_H^2}. \tag{4.58}$$

For gauge couplings $g_1 \sim 0.35$, $g_2 \sim 0.65$ we have $E = 0.0067$ which means we need the tree-level Higgs mass to be about a twelfth of the vacuum expectation value to obtain an order parameter of unity. Clearly the Higgs mass is far too heavy to accommodate this. If the electroweak phase transition is strongly first-order we need extra bosons to couple to the Higgs.

4.3.2 Issues of gauge invariance

So far our analysis of the Standard Model effective potential completely ignores gauge invariance. Gauge dependence manifests itself at the one-loop level since the propagators are gauge-dependent [17]. This gauge dependence is directly due to the masses of Goldstones and ghosts which are gauge dependent and therefore appears in the one-loop potential. The Higgs field has four degrees of freedom which we can put into a column vector $\phi = (\phi_1, \phi_2, \phi_3, \phi_4)$ along with the generating function $j = j_i$ to simplify notation. At this stage to make the analysis as general as possible we will consider a model with a scalar field that couples to a non-abelian gauge field. The gauge fixed generating functional for such a model is

$$Z\,[j] = \int \mathcal{D}\phi \mathcal{D}A \mathcal{D}\eta \mathcal{D}\eta^\dagger e^{i\int d^4x(\mathcal{L}(x,\, j,\, \xi))} \tag{4.59}$$

with

$$\begin{aligned}\mathcal{L}(x, j, \xi) = &- V(\phi_{\rm cl}) + j\phi_{\rm cl} + \phi_i\left(-\frac{\partial V}{\partial \phi} + j_i\right)\\ &+ \frac{1}{2}\phi_i\left[-\partial^2 - M_{ij}^2(\phi_{\rm cl}) - \xi m_A^2(\phi_{\rm cl})_{ij}\right]\phi_j\\ &+ \frac{1}{2}A_\mu^a\left[(\partial^2 g^{\mu\nu}) - \left(1 - \frac{1}{\xi}\partial^\mu\partial^\nu\right)\delta^{ab} + m_A^2(\phi_{\rm cl})^{ab} g^{\mu\nu}\right]A_\nu^a\\ &+ \eta^{\dagger a}\left[-\partial^2\delta^{ab} - \xi m_A^2(\phi_{\rm cl})^{ab}\right]\eta^b\end{aligned} \tag{4.60}$$

and the field-dependent masses are defined

$$\begin{aligned}M_{ij}^2(\phi_{\rm cl}) &= \left.\frac{\partial^2 V}{\partial\phi_i\partial\phi_j}\right|_{\phi_{\rm cl}}\\ m_A^2(\phi_{\rm cl})^{ab} &= \left(gT^a\phi_{\rm cl}\right)_i\left(gT^b\phi_{\rm cl}\right)_j\\ \xi m_A^2(\phi_{\rm cl})_{ij} &= \xi m_A^2(\phi_{\rm cl})^{aa}.\end{aligned} \tag{4.61}$$

Following the same steps as outlined in the previous sections, one finds that the effective potential to one loop, ignoring fermions, is

$$
\begin{aligned}
V = V_{\text{tree}}(\phi) &+ \sum_{\text{scalars},i} \frac{[m_i^2(\phi, \xi)]^2}{64\pi^2}\left[\log\left(\frac{m_i^2(\phi, \xi)}{\mu^2}\right) - \frac{3}{2}\right] \\
&+ 3\sum_{\text{gauge},a} \frac{[m_a^2(\phi)]^2}{64\pi^2}\left[\log\left(\frac{m_a^2(\phi)}{\mu^2}\right) - \frac{5}{6}\right] - \sum_{\text{gauge},a} \frac{[\xi m_a^2(\phi)]^2}{64\pi^2} \\
&\left[\log\left(\frac{\xi m_a^2(\phi)}{\mu^2}\right) - \frac{3}{2}\right] + \frac{T^4}{2\pi^2}\left[\sum_{\text{scalar},i} J_B\left(\frac{m_i^2(\phi, \xi)}{T^2}\right)\right. \\
&\left. + 3\sum_{\text{gauge},a} J_B\left(\frac{m_a^2(\phi)}{T^2}\right) - \sum_{\text{gauge},a} J_B\left(\frac{\xi m_a^2(\phi)}{T^2}\right)\right].
\end{aligned}
\tag{4.62}
$$

Making the same approximations as before one finds that the order parameter is gauge-dependent. Fortunately, the leading order term, $O(\xi^{3/2})$ arises from the Goldstones and ghosts with opposite signs and cancels for large values of the gauge fixing parameter. Even still, it is morally disturbing that the key parameter in our problem is gauge dependent.

All physical properties have to be gauge-independent by definition and the order parameter is indeed a physical quantity—the point where any reduction in temperature makes it energetically favorable to tunnel from the false vacuum to the true vacuum. So if we find the order parameter has become gauge-dependent then it must be an artifact of our approximation scheme in expanding the effective action at finite temperature to one loop and finding the minima.

In order to calculate a gauge-independent order parameter we therefore need to find a gauge-independent expansion of v_c and T_C or $V_{\text{eff}}(T)$. Such a scheme is known as the hbar expansion, and is best performed after dimensionally reducing in order to properly handle issues of the convergence of perturbation theory. We will explain this in chapter 9, but for now we will follow reference [17] in doing a hbar expansion in the unresummed four-dimensional theory to conceptually explain the approach. Let us begin with the Nielsen identity [18, 19],

$$
\frac{\partial V_{\text{eff}}}{\partial \xi} = -C_i(\phi, \xi)\frac{\partial V_{\text{eff}}}{\partial \phi_i}, \tag{4.63}
$$

which means the effective potential is gauge-independent at its extrema. However, the extrema at the one-loop level are gauge-dependent. Since the extrema of the full effective potential must be gauge-independent (as it is a physical quantity) again the gauge dependence of the minimum must be an artifact of our approximation scheme.

Let us expand both C_i and the effective potential as a series in $\hbar$. That is $V_{\text{eff}} = V_0(\phi) + \hbar V_1(\phi) + \cdots$ and $C = c_0 + \hbar c_1 + \cdots$, and V_1 is of course the one-loop correction to the effective potential. Substituting this into the Nielsen identity we obtain

$$\frac{\partial V_0}{\partial \xi} + \hbar \frac{\partial V_1}{\partial \xi} = -c_0 \frac{\partial V_0}{\partial \phi} - \hbar \left(c_0 \frac{\partial V_1}{\partial \phi} + c_1 \frac{\partial V_0}{\partial \phi} \right). \tag{4.64}$$

We then equate terms of each power of $\hbar$ on either side to each other. The zeroth-order terms are straightforward to handle since the tree-level effective potential is manifestly gauge invariant which implies $c_0 = 0$. For terms linear in $\hbar$ we obtain

$$\frac{\partial V_1}{\partial \xi} = -c_1 \frac{\partial V_0}{\partial \phi}, \tag{4.65}$$

which implies that the one-loop correction to the effective potential is gauge invariant at the tree-level minima, and *not* the one-loop minima. Let us make a $\hbar$ expansion to the minima as previously promised, $\phi_c = \phi_0 + \hbar \phi_1 + \cdots$, where ϕ_0 are the tree-level minima. It is then straightforward to find gauge invariant corrections to the tree-level minima as follows

$$\begin{aligned} \left.\frac{\partial V_{\text{eff}}}{\partial \phi}\right|_{\phi_c} = 0 = \left.\frac{\partial V_0}{\partial \phi}\right|_{\phi_0 + \hbar\phi_1 + \cdots} + \hbar \left.\frac{\partial V_1}{\partial_\phi}\right|_{\phi_0 + \hbar\phi + \cdots} \\ = \left.\frac{\partial V_0}{\partial \phi}\right|_{\phi_0} + \left.\left[\frac{\partial V_0}{\partial \phi} + \hbar\left(\frac{\partial V_1}{\partial \phi} + \phi_1 \frac{\partial^2 V_0}{\partial \phi^2}\right)\right]\right|_{\phi=\phi_0} + \cdots. \end{aligned} \tag{4.66}$$

Insisting that each power of $\hbar$ vanishes gives a gauge invariant expression for ϕ_1

$$\phi_1 = -\left.\left[\left(\frac{\partial^2 V_0}{\partial \phi^2}\right)^{-1} \frac{\partial V_1}{\partial \phi}\right]\right|_{\phi=\phi_0}. \tag{4.67}$$

Substituting this back into our $\hbar$ expansion for the effective potential we can state a condition for the critical temperature which is valid to second order in our expansion

$$\begin{aligned} V(\phi_c, T_C) = V_0(\phi_0) + \hbar V_1(\phi_0, T_C) \\ + \hbar^2 \left[V_2(\phi_0, T_C) - \frac{1}{2}\phi_1(\phi_0)^2 \left.\frac{\partial V_0}{\partial \phi}\right|_{\phi_0} \right] + \cdots = 0. \end{aligned} \tag{4.68}$$

The critical temperature is then evaluated by solving the above equation.

4.3.3 The sphaleron rate at high temperature

Now that we have the necessary knowledge of both finite temperature QFT and the evolution of the vacuum expectation value with temperature, we can fill in the gaps we previously left in the calculation of the sphaleron rate at high temperature [22, 23]. If one assumes that the sphaleron gas is very dilute we can expand around the single sphaleron solution. We will take the sphaleron rate to be the probability of finding a right moving particle at the barrier ($x = 0$) of the periodic potential times the rate at which it crosses the barrier. If the temperature is very large then it is justified to work in a classical regime where we simultaneously define position and momentum. This means $\Gamma = \langle \delta(x)\Theta(p) \rangle$ as the step function ensures a right-moving particle at a barrier which is at the origin. The classical rate is

$$\Gamma = \frac{\int \mathrm{d}p\mathrm{d}x \, \exp\left[-\beta\left(\frac{1}{2}p^2\right) + V(x)\right]\delta(x)\Theta(p)}{\int \mathrm{d}p\mathrm{d}x \, \exp\left[-\beta\left(\frac{1}{2}p^2\right) + V(x)\right]} \approx \frac{\omega_0}{2\pi}\mathrm{e}^{-\beta V_0}. \tag{4.69}$$

We can relate this to the imaginary part of the free energy of a single sphaleron with small fluctuations

$$\begin{aligned} \mathrm{Im}\, F = T\, \mathrm{Im} \ln Z &= T\frac{\mathrm{Im}\, Z}{Z} \\ &\approx T\frac{\mathrm{Im} \int \mathrm{d}p\mathrm{d}x \, \exp\left[-\beta\left(\frac{p^2}{2} + V_0 - \frac{1}{2}\omega x^2\right)\right]}{\int \mathrm{d}p\mathrm{d}x \, \exp\left[-\beta\left(\frac{p^2}{2} + \frac{1}{2}\omega x^2\right)\right]} \\ &= \frac{\omega_0}{2\omega - \beta}\mathrm{e}^{-\beta V_0}, \end{aligned} \tag{4.70}$$

which means we can write

$$\Gamma \approx \frac{\omega_\beta}{\pi}\frac{\mathrm{Im}\, Z_{\mathrm{sph}}}{Z_0}. \tag{4.71}$$

Zero modes can be integrated out and each give a factor of NV where V is the volume of the symmetry group and N is a normalization factor. The sphaleron solution derived before has a $SU(2)_L$ spherical and translational symmetry so we obtain a factor of $(\mathcal{NV})_{\mathrm{tr}}$ as well as a factor of $(\mathcal{NV})_{\mathrm{rot}}$. Defining the coupling constant of the finite temperature electroweak theory as $g_3 = (g^2 T/2M_W)^{1/2}$, the sphaleron rate is

$$\begin{aligned} \Gamma &\approx \frac{\omega_-}{2\pi}(\mathcal{NV})_{\mathrm{rot}}(\mathcal{NV})_{\mathrm{tr}}\, \mathrm{Im}\left[\frac{\det g_3^{-2}\Omega_0^2}{\det' g_3^{-2}\Omega_{\mathrm{sph}}^2}\right]^{1/2} \mathrm{e}^{-E_{\mathrm{sph}}/T} \\ &= \frac{\omega_-}{2\pi}(\mathcal{NV})_{\mathrm{rot}}(\mathcal{NV})_{\mathrm{tr}}\, \mathrm{Im}\left[\frac{\det \Omega_0^2}{\det' \Omega_{\mathrm{sph}}^2}\right]^{1/2} g_1^{-N_0}\mathrm{e}^{-E_{\mathrm{sph}}/T} \\ &\equiv \frac{\omega_-}{2\pi}(\mathcal{NV})_{\mathrm{rot}}(\mathcal{NV})_{\mathrm{tr}}\kappa g_3^{-N_0}\mathrm{e}^{-E_{\mathrm{sph}}/T} \\ &= \frac{\omega_-}{2\pi}(\mathcal{NV})_{\mathrm{rot}}(\mathcal{NV})_{\mathrm{tr}}\kappa\left(\frac{v(T)}{gT}\right)^3 \mathrm{e}^{-E_{\mathrm{sph}}/T}. \end{aligned} \tag{4.72}$$

In the above we have implicitly defined the fluctuation determinant as $\kappa \approx 20$ for notational simplicity. As we derived before the sphaleron energy is given by

$$S_3[\phi_{\rm sph}] \equiv \frac{E_{\rm sph}}{T} = \frac{4\pi v(T)}{gT} B\left(\frac{\lambda}{g^2}\right) \tag{4.73}$$

with

$$B\left(\frac{\lambda}{g^2}\right) = \int_0^\infty dr \left[4\left(\frac{df}{dr}\right)^2 + \frac{8}{r^2} f^2 (1-f)^2 + \frac{r^2}{2}\left(\frac{dh}{dr}\right)^2 h^2(1-f)^2 + \frac{1}{4}\frac{\lambda}{g^2} r^2 (1-h^2)^2\right], \tag{4.74}$$

where $r \equiv gv(T)|\, x\, |^2$. In the symmetric phase the sphaleron rate can be determined from dimensional analysis with the only relevant scale being the thermal mass of the gauge bosons gT^2. Specifically, the sphaleron rate is

$$\frac{\Gamma_{\rm sph}}{V} = 6\kappa\alpha_W^5 T. \tag{4.75}$$

Note that if there is a discontinuous jump in $v(T)$ as happens in a strongly first-order phase transition, the sphaleron rate can be much smaller than the Hubble rate, $H \approx \sqrt{g^*}\, T^2/M_{\rm pl}$, inside a bubble of broken phase, while being fast outside such a bubble. In figure 4.8 we show a schematic of the sphaleron rate compared to the Hubble rate turning a strongly first-order phase transition.

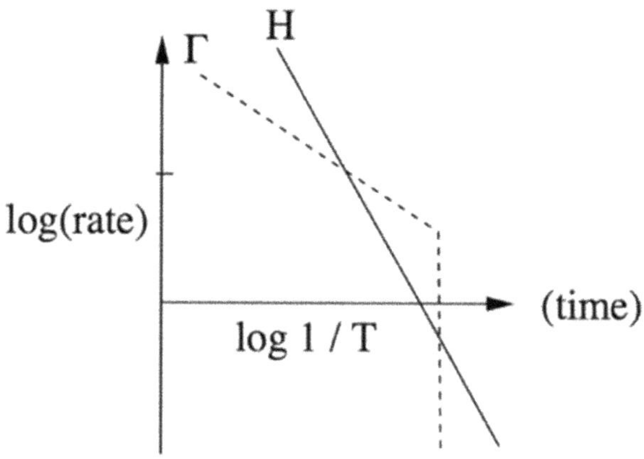

Figure 4.8. The sphaleron rate at high temperature in the advent of a strongly first-order phase transition. If the electroweak phase transition is sufficiently strongly first-order weak sphalerons drop out of equilibrium being fast compared to the Hubble rate outside the bubble and slow inside. Reproduced from [12].

4.4 The bounce solution

The transition from the high temperature minimum to the true vacuum occurs via tunneling processes. We would like to understand the field profile that describes this tunneling process. The action has a local extremum known as the bounce solution where the fields continuously vary from one vacuum to the other only just tunneling through the barrier separating the two minima [25].

Recall that the Euclidean action is

$$S_E = \int d^d x \left[\frac{1}{2} (\partial_\mu \phi_i)^2 + V(\phi_i) \right], \tag{4.76}$$

with $d = 3$ for finite temperature. The nucleation rate per unit volume is

$$\Gamma = A T^4 e^{-\frac{S_E}{T}} \tag{4.77}$$

[26, 27], where T is the temperature, A is the fluctuation determinant, and S_E is the Euclidean action for the bounce solution which satisfies the classical equations of motion. Numerically, setting the expectation value for the number of bubbles per unit volume to unity, one finds $S_E/T \approx 140$ [28, 29] if the temperature is at the electroweak scale. More generally, one finds $\frac{S_E}{T} = 170–4 \ \text{In}\left(\frac{T_N}{1\text{GeV}}\right)^{-2} \ \text{In} \ g_*$. Assuming a spherically symmetric bubble profile we can write the classical equations of motion in terms of a single space–time variable $\rho \equiv \sqrt{r^2 - t^2}$,

$$\frac{\partial^2 \phi_i}{\partial \rho^2} + \frac{(d-1)}{\rho} \frac{\partial \phi_i}{\partial \rho} - \frac{\partial V}{\partial \phi_i} = 0, \tag{4.78}$$

and the bounce solution satisfies the conditions $\phi_i(0) \approx v_i^{\text{true}}$, $\phi_i(\infty) = v_i^{\text{false}}$ and $\phi_i'(0) = 0$[3]. Here ρ^2 is the space–time coordinate and v_i^{false} and v_i^{true} are the vacuum expectation values (VEVs) of the field ϕ_i in the true and false vacua, respectively. The classical equations of motion imitate a ball rolling in a landscape of shape $-V$ with ρ playing the role of time, but including a ρ-dependent friction term.

Using this analogy of a ball in a landscape, the true and false vacua are global and local maxima to the landscape, respectively, and they are separated by a valley. The solution that just pierces the barrier via tunneling processes is analogous to the ball starting from rest near the top of the global maximum and being released such that it comes to rest at $\rho = \infty$ with the ball settling on the local maximum. The time-dependent friction term means the ball loses energy along the journey so it must start at a point higher than the local maximum. If the starting point is too low it will come to rest at the bottom of the valley as $\rho \to \infty$. Alternatively if the starting point is too high the ball will go past the local maximum and diverge. The bounce solution is a fine-tuned point just between these two extremes and the total action (classically the action is approximately equal to energy times time) of the path that differs slightly from the bounce is clearly higher. So it is a local minimum.

[3] The first is not literally a boundary condition, unlike the other two.

4.5 Analytic techniques for the single-field case

4.5.1 Single-field approximation

Calculating the bounce solution for a generic multi-field case is a highly non-trivial task. Generally, it is easier to approximate the system by a simpler system then use numerical or analytic techniques to improve the approximation [30, 31]. The simplified system that is generally used is a single-field problem.

To approximate the solution to the multi-field problem with a single-field solution, the steps are as follows. Begin by translating the effective potential such that the false vacuum is at the origin in field space. Then perform a rotation so that the true vacuum lies along one of the axes in the rotated frame in field space, say $\hat{\phi}_1$. Rewrite the effective potential in the rotated basis $V(\varphi_1, \varphi_2, \ldots) \to V(\phi_1, \phi_2, \ldots)$. Finally, one only considers the shape of the potential along the $\hat{\phi}_1$ direction by making the transformation

$$V(\phi_1, \phi_2, \ldots) \to V(\phi_1, 0, \ldots). \tag{4.79}$$

The equation of motion for the simplified potential is much more tractable

$$\frac{\partial^2 \phi_1}{\partial \rho^2} + \frac{(d-1)}{\rho}\frac{\partial \phi_1}{\partial \rho} - \frac{\partial V(\phi_1, 0, \ldots)}{\partial \phi_1} = 0.$$

The solution to this provides an ansatz that can be perturbed around to find the solution for the multi-field case.

Any system with three turning points where a local and global minimum are separated by a barrier can be approximated by the most general renormalizable potential with a single field

$$V(\varphi) = M^2\varphi^2 + b\varphi^3 + \lambda\varphi^4.$$

This includes the complicated curves that result from the complicated thermal functions[4]. Although it looks like there are three independent parameters, two can just be absorbed into rescaling transformations. For example, if we make the rescaling $\phi = \varphi_{\min}\ \varphi$, where $\varphi_{\min}$ is the global minimum, the rescaled potential will have its global minimum at $\phi = 1$. Such a rescaling needs to be compensated by a rescaling in the spatial parameters so that the effective action remains unchanged. This is achieved by the replacement $\rho \to \varphi_{\min}\,\rho$. A convenient parametrization for the rescaled potential is [32][5]

$$V(\phi) = \frac{(4\alpha - 3)}{2}E\phi^2 + E\phi^3 - \alpha E\phi^4. \tag{4.80}$$

The strength of such a parametrization is that the second rescaling—stretching or contracting the V axis—is manifest by changing the value of E. Let us assume that the three turning points are in the positive ϕ direction (the other case is covered by reflection transformation). A barrier between the local and global maxima requires that $E < 0$ and $\alpha \in (0.5, 0.75)$. As shown in figure 4.9, increasing the value of α reduces the height of the barrier separating the minimum from the extreme cases of

[4] An approximate expression for the effective action of a similar potential can be found in [32].
[5] This definition of α differs from that of [33] but the physical principles are the same.

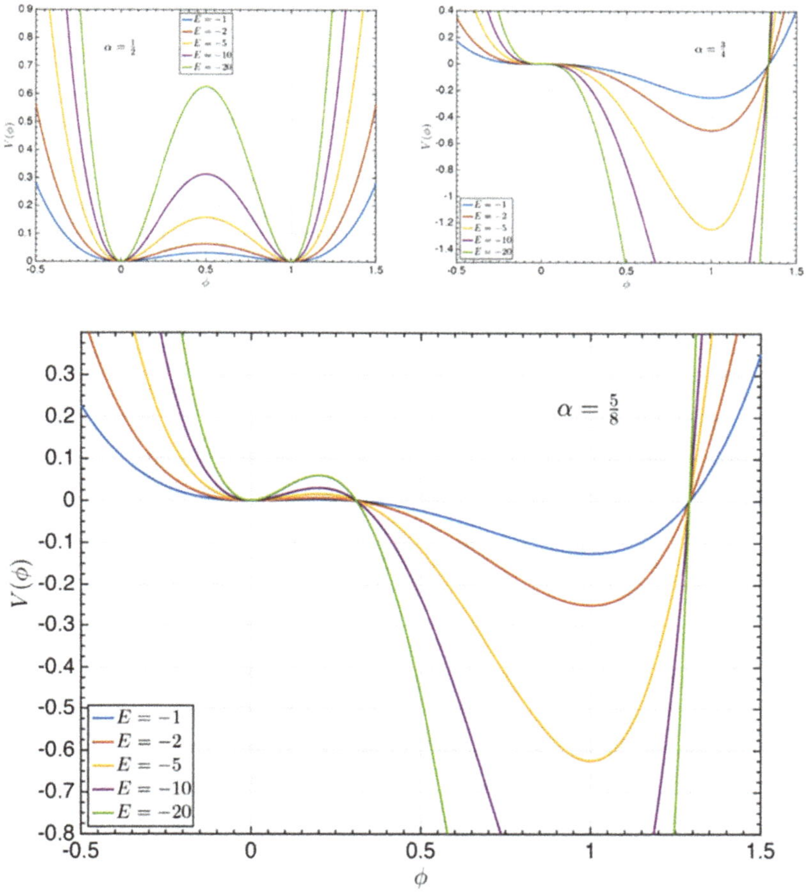

Figure 4.9. The most general tree-level renormalized potential rescaled such that the minimum is at one. The potential is parametrized by α and E which are given in equation (4.80). Every panel has the potential drawn for $-E = 1,2,5,10,20$. The parameter $\alpha \in [0.5, 0.75]$ is in a compact space. The larger panel has the mean value in $\alpha = \frac{5}{8}$, whereas the other two have the two edge cases with $\alpha = \frac{1}{2}$ on the left and $\alpha = \frac{3}{4}$ on the right. Figure taken from reference [32].

$\alpha = \frac{1}{2}$ where the minima are degenerate to $\alpha = \frac{3}{4}$ where the barrier disappears. Increasing $|E|$ increases the overall scale of the potential.

4.5.2 Developing ansatz solutions

We can derive an approximate bounce solution to all potentials of the form given in equation (4.80) as it depends only on a single compact parameter, α. The parameter E is just a scale factor. Thus, one can factor $|E|$ out of the equations of motion by further rescaling $\rho \to \rho/\sqrt{|E|}$. Then, the equations of motion only depend on α.

Explicitly our rescaling operations are,

$$\begin{Bmatrix} \varphi \to \phi = \varphi_{\min} \;\; \varphi \\ \rho \to \frac{\varphi_{\min}}{\sqrt{|E|}}\rho \end{Bmatrix}$$

under which the Euclidean action becomes

$$S_E = 4\pi\frac{\phi_m^3}{\sqrt{|E|}}\int \mathrm{d}\rho\,\rho^2\left[\left(\frac{\partial\phi}{\partial\rho}\right)^2 - \tilde{V}(\phi)\right]$$

where[6] $\tilde{V} \equiv V/|E|$, and we have assumed isotropy which allows us to integrate over the solid angle. The effective action is now proportional to an integral that only depends on α, that is

$$S_E = 4\pi\frac{\phi_m^3}{\sqrt{|E|}}f(\alpha). \tag{4.81}$$

The bounces solutions can be found for the full parameter space via an undershoot overshoot method where one integrates the equation of motion numerically for various starting points until one finds the bounce, as described in the previous section. These solutions can be approximated by kink solutions [34]

$$\phi \approx \frac{1}{2}\left(1 - \tanh\left[\frac{\rho - \delta(\alpha)}{L_w(\alpha)}\right]\right), \tag{4.82}$$

with parameters δ and L_w describing the offset and the bubble wall width, respectively. If one curve fits the bounce solution to the kink solution numerically over the full parameter space, $\delta(\alpha)$ and $L_w(\alpha)$ become α-dependent functions. Also, the integral in the Euclidean action, $f(\alpha)$, becomes α-dependent.

Some quick curve fitting over the entire parameter space gives the following functions for $\delta(\alpha)$, $L_w(\alpha)$ and $f(\alpha)$, respectively,

$$\delta(\alpha) \approx \delta_0 + k\left(\alpha - \frac{5}{8}\right) + \sum_{n=1}^{2} a_n\left[\frac{\alpha - \frac{5}{8}}{\left(\alpha - \frac{1}{2}\right)\left(\alpha - \frac{3}{4}\right)}\right]^{(2n-1)}$$

$$L_w(\alpha) \approx \mathcal{L}_0\left[\left(\alpha - \frac{1}{2}\right)^r + \frac{c}{|\alpha - \frac{3}{4}|^s}\right]$$

and

$$f(\alpha) = f_0\frac{|\alpha - \frac{3}{4}|^p}{|\alpha - \frac{1}{2}|^q}.$$

These functions are in excellent agreement with the numerical solution for the values given in table 4.1, which can be seen in figure 4.10.

Let us conclude with a note on how the tunneling rate depends on the remaining parameters of the single-field potential. First note that $|E|$ scales as $|b\phi^3|$.

[6] V does not have any dependence on $|E|$.

Table 4.1. Table of fitter values that parametrize the functions that approximate the bubble wall width L_w, and the offset δ from the kink solution, as well as f which parametrizes the Euclidean action. Table reproduced from [32].

$f(\alpha)$		$L_w(\alpha)$		$\delta(\alpha)$	
Parameter	**Value**	**Parameter**	**Value**	**Parameter**	**Value**
$f0$	0.0871	ℓ_0	1.4833	δ_0	2.2807
p	1.8335	c	0.4653	k	−4.6187
q	3.1416	r	18.0000	a_1	0.5211
		s	0.7035	$a_2 \times 10^5$	7.8756

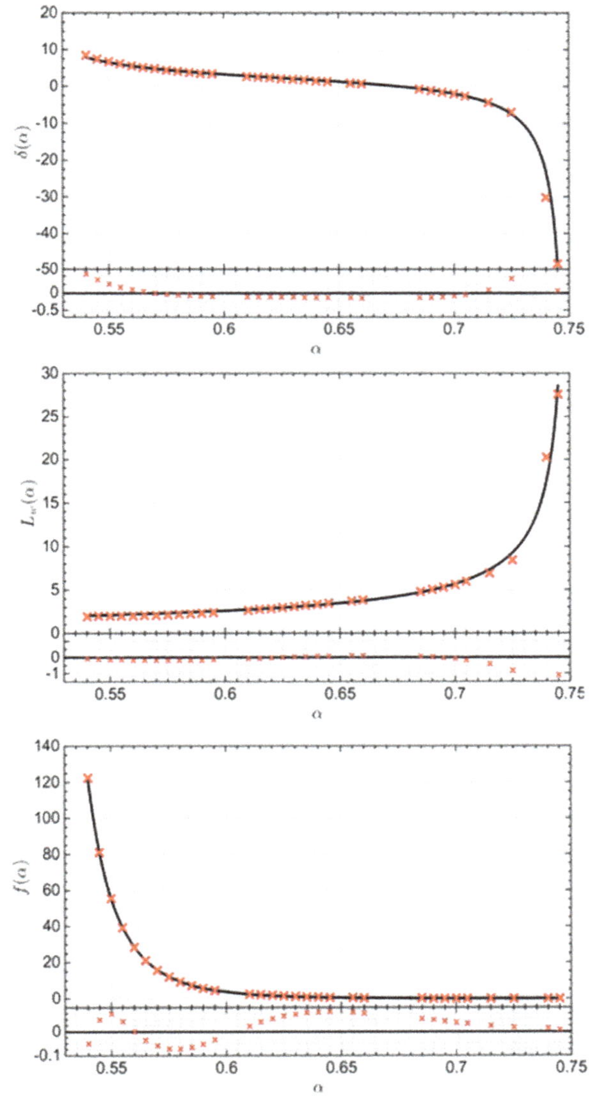

Figure 4.10. The fits to the bubble wall width L_w (top panel) and the offset δ (middle panel) of the kink solution, as well as the integral f (lowest), which appears in the Euclidean action. Both the numerically computed values and the residuals are given. Reproduced from [32].

Therefore, S_E/T scales as $\frac{\phi_m}{T}\sqrt{\frac{\phi_m}{b}}$, where b is the cubic coupling of the unscaled potential given in section 4.5.1 and therefore controls the height of the barrier.

4.6 Path deformation method

Any one-dimensional problem can be approximated fairly well by the set of kink solutions given in the previous section. One might ask if it is also possible in general to write an ansatz for the multi-field problem. Indeed, such a method was attempted by [34] for a series of parametrized ansatzes

$$\phi_i(\rho) = \frac{\phi_{0i}}{2}\left(1 - \tanh\left[\frac{\rho + \delta_i}{L_i}\right]\right). \tag{4.83}$$

One can then in principle minimize the effective action with respect to the parameters $(\phi_{0i}, \delta_i, L_i)$. Unfortunately, this method turns out to be virtually intractable numerically for a lot of potentials. The reason is that for particularly large values of $| \delta_i |$ or L_i the solution begins to resemble the solution of a constant field in a global minimum. As such some more sophisticated methods have been explored. We will give a detailed explanation of a semi-analytic method but first let us review a nice numerical method that involves deforming a path. This method takes advantage of the fact that you can always solve the one-dimensional solution through bisection. Therefore, let us parametrize the path through field space which the tunneling solution follows by x. We can then separate the classical equations of motion into an equation along the path x and the orthogonal directions

$$\begin{aligned} \frac{\mathrm{d}^2 x}{\mathrm{d}\rho^2} + \frac{2}{\rho}\frac{\mathrm{d}x}{\mathrm{d}\rho} &= \frac{\partial V}{\partial x} \\ \frac{\partial^2 \vec{\phi}}{\mathrm{d}x^2}\left(\frac{\mathrm{d}x}{\mathrm{d}\rho}\right)^2 &= \nabla_\perp V. \end{aligned} \tag{4.84}$$

One can solve the first equation by bisecting the starting position x_0 along any given path and then deform the path by making steps in the orthogonal directions until the second set of equations is satisfied. Once the new path is defined one can solve the first equation along the new path. These two steps are repeated until convergence is achieved. One drawback to this method is that if the steps used to deform the path are too large then convergence might not be achieved. Nonetheless it is an elegant numerical solution that has been implemented in a publicly available package [31].

4.7 Perturbative method

Another method for finding the bubble wall profile can be found in [32] and involves finding an ansatz which is near to the true solution and perturbing around it. The key assumption is to assume that the trajectory in field space of the bounce solution does not differ from the single-field approximate trajectory by an amount that is greater than the distance between the two minima. In other words, one assumes that

$$| \phi_i(\rho)| \lesssim | v_{\rm true} - v_{\rm false} |.$$

For the rescalings derived in section 4.5, the maximum distance between the ansatz and true solution is bounded by 1, although in practice it is usually much smaller. We will denote the ansatz solution for the field ϕ_i as A_i with perturbations denoted as ϵ_i. That is, $\phi_i = A_i + \epsilon_i$. Inserting this into the classical equations of motion (4.84) yields

$$\frac{\partial^2 A_i}{\partial \rho^2} + \frac{\partial^2 \epsilon_i}{\partial \rho^2} + \frac{2}{\rho}\frac{\partial A_i}{\partial \rho} + \frac{2}{\rho}\frac{\partial \epsilon_i}{\partial \rho} = \left.\frac{\partial V(\phi)}{\partial \phi_i}\right|_A + \sum_j \left.\frac{\partial^2 V(\phi)}{\partial \phi_i \partial \phi_j}\right|_A \epsilon_j,$$

where $A \equiv \{A_i\}$. The above equation can be organized into terms involving the ansatz and terms involving the corrections. The result is a network of coupled inhomogeneous differential equations for the perturbations ϵ_i:

$$\frac{\partial^2 \epsilon_i}{\partial \rho^2} + \frac{2}{\rho}\frac{\partial \epsilon_i}{\partial \rho} - \left.\frac{\partial^2 V(\phi)}{\partial \phi_i \partial \phi_j}\right|_{A_j} \epsilon_i = B_i(\rho).$$

We have implicitly defined the inhomogeneous pieces of the differential equations as $B_i(\rho)$. These functions are given by

$$B_i(\rho) \equiv \left.\frac{\partial V(\phi)}{\partial \phi_i}\right|_A - \frac{\partial^2 A_i}{\partial \rho^2} - \frac{2}{\rho}\frac{\partial A_i}{\partial \rho}.$$

Clearly the function B_i is a measure of how well the ansatz solves the equations of motion, as one can see that if B_i is zero then the corrections to the ansatz are defined by a set of homogeneous differential equations that are satisfied by the trivial solution $\epsilon_i = 0$.

A key observation of [32] is that one can linearize and approximately solve these differential equations analytically by approximating the mass matrix by a series of step functions, with a correction which can be used to form the convergent series of perturbations. The techniques used to derive the solution are given in section 8.3. Here we just use the easier parametrized ansatz approach from that section where the form of the solution is assumed up to certain parameters and substituted back into the differential equations to determine such parameters.

To make use of the parametrized ansatz approach, consider the homogeneous form ($Bi = 0$) of the differential equations for ϵ. The solutions take the form

$$\epsilon \sim \frac{z e^{\lambda \rho}}{\rho}.$$

Inserting this into the set of differential equations results in an eigenvalue problem with the mass matrix. We parametrize the homogeneous solutions with the following notation

$$\epsilon_{ik}^h = \frac{z_{ik} e^{\lambda_k \rho}}{\rho}.$$

Here k spans the n eigenvalues of the mass matrix and is not summed over, whereas i is the field index. Substituting ϵ_{ik}^{h} for ϵ_i into the homogeneous form of the differential equations gives

$$\sum_j M_{ij} z_{jk} = \lambda_k^2 z_{ik},$$

with M being the mass matrix

$$M_{ij} = \left.\frac{\partial^2 V(\phi)}{\partial\phi_i \partial\phi_j}\right|_A .$$

Note here that z_{ik} is the ith component for the eigenvector of the mass matrix with eigenvalue λ_k^2. Since the differential equations are second order, one finds that λ^2 are the eigenvalues of the mass matrix so the set of homogeneous solutions involve both the positive and negative roots, $\pm\lambda_k$. We therefore introduce $\tilde{\lambda}_j$ and $\tilde{z}_{ij}$, where

$$\tilde{\lambda}_1 = \lambda_1,\ \tilde{\lambda}_2 = -\lambda_1,\ \tilde{\lambda}_3 = \lambda_2,\ \ldots \tag{4.85}$$

$$\tilde{z}_{i1} = z_{i1},\ \tilde{z}_{i2} = z_{i1},\ \tilde{z}_{i3} = z_{i2},\ \tilde{z}_{i4} = z_{i2},\ \ldots \tag{4.86}$$

such that in $\tilde{\lambda}_j$ and $\tilde{z}_{ij}$, the index j spans the range 1,2, … ,2n. Using the techniques described in [33] one can derive that the particular solution is

$$\epsilon_i^{\lessgtr} = \sum_{j=1}^{2n}\sum_{k=1}^{n} \tilde{z}_{ij} \frac{e^{\tilde{\lambda}_j \rho}}{\rho}\left(\int_0^{\rho} dt t e^{-\tilde{\lambda}_j t} h_{jk} B_k^{\lessgtr}(t) - \beta_j^{\lessgtr}\right).$$

Here the constants $\beta_j^{\lessgtr}$ are determined by the boundary and matching conditions and the inhomogeneous functions are denoted by B_k, with k the equation number. To determine the value of the weighting constants h_{jk} one needs to invert the $2n^2$ constraint equations

$$\sum_{j=1}^{2n} \tilde{z}_{ij} h_{jk} \tilde{\lambda}_j = \delta_{ik} \tag{4.87}$$

$$\sum_{j=1}^{2n} \tilde{z}_{ij} h_{ik} = 0. \tag{4.88}$$

The above equations are all matrix equations of rank n which result in $2n^2$ constraints in total. This of course is the solution for the case where the mass matrix is approximated by a series of step functions. If we wish to handle its full space–time varying profile, this can be achieved by treating corrections to the above solution also as a perturbation. Let us relabel the above solution ϵ_i^0 and substitute $\epsilon_i = \epsilon_i^0 + \delta\epsilon_i + \cdots$ into the differential equations denoting the deviation from the step function approximation of the mass matrix as $\eta_{ij}(\rho)$. If we keep only first-order terms we once again find we have a linear set of differential equations

$$\left.\frac{\partial^2 \epsilon_i}{\partial \rho^2} + \frac{2}{\rho}\frac{\partial \epsilon_i}{\partial \rho} - \frac{\partial^2 V(\phi)}{\partial \phi_i \partial \phi_j}\right|_A \epsilon_j = B_i(\rho) \frac{\partial^2 \delta\epsilon_i + \epsilon_i^0}{\partial \rho^2} + \frac{2}{\rho}\frac{\partial \delta\epsilon_i + \epsilon_i^0}{\partial \rho} - \left(\delta\epsilon_j + \epsilon_j^0\right)\left(\bar{M}_{ij} + \eta_{ij}\right) = B_i(\rho). \tag{4.89}$$

We can observe that ϵ_i^0, by definition, is the solution to the initial differential equations with mass matrix $\bar{M}_{ij}$. This allows a dramatic simplification in the above equation

$$\begin{aligned}\frac{\partial^2 \delta\epsilon_i}{\partial \rho^2} + \frac{2}{\rho}\frac{\partial \delta\epsilon_i}{\partial \rho} - \delta\epsilon_j \bar{M}_{ij} - \eta_{ij}\epsilon_j^0 = 0 \\ \frac{\partial^2 \delta\epsilon_i}{\partial \rho^2} + \frac{2}{\rho}\frac{\partial \delta\epsilon_i}{\partial \rho} - \delta\epsilon_j \bar{M}_{ij} = \eta_{ij}\epsilon_j^0.\end{aligned} \tag{4.90}$$

Once again we have a network of coupled linear differential equations which has the same form, and therefore the same form of the solution, as the differential equations for ϵ_i^0. Therefore, to write the solution to the above equation we merely replace B_i with $\eta_{ij}\epsilon_j^0$. To obtain the bounce solution, simply calculate the series ϵ_i with corrections $\delta\epsilon_i$ up to a desired tolerance.

4.7.1 Observations on convergence

The semi-analytic approach has similarities to a functional version of Newton's method. This raises the question of whether it might face analogous convergence issues. That is:

1. Will convergence be slow if the guess is far from the true solution?
2. Are there cases where the perturbations oscillate rather than converge? That is, can $\epsilon^{(n)}(\rho) \approx -\epsilon^{(n+1)}(\rho)$?
3. Is there an analogous problem to where the perturbation diverges like it does in Newton's method when the derivative is zero?
4. Can you converge toward the wrong solution due to being in the wrong basin of attraction?

The semi-analytic perturbative method is immune from any of these problems, except the fourth, which is a general problem for all known algorithms which attempt to find bubble wall profiles. We previously made an argument that this method is generically free from the first problem as the choice of the initial ansatz generically ensures perturbations bounded by 1. The remaining two issues are more difficult to deal with.

We begin by discussing the possibility of oscillating solutions. Let the ansatz with an n perturbative correction be denoted by

$$A_i^{(n)} = A_i^{(0)} + \sum_{k=1}^{n-1} \epsilon_i^{(k)}.$$

In the above, we denote the initial ansatz as $A_i^{(0)}$ with corrections $\epsilon_i^{(k)}$. Let us assume that for a particular field ϕ_i, each correction oscillates at correction n, such that $\epsilon_i^{(n)} = -\epsilon_i^{(n+1)}$. We can then write the equations of motion for $A_i^{(n+2)} = A_i^{(n)} + \epsilon_i^{(n)} + \epsilon_i^{(n+1)}$ before Taylor expanding as

$$\begin{aligned} &\frac{\partial^2[A_i^{(n)} + \epsilon_i^{(n)} + \epsilon_i^{(n+1)}]}{\partial\rho^2} + \frac{2}{\rho}\frac{\partial[A_i^{(n)} + \epsilon_i^{(n)} + \epsilon_i^{(n+1)}]}{\partial\rho} + \left.\frac{\partial V(\phi)}{\partial\phi_i}\right|_{\left\{A_k^{(n)}+\epsilon_k^{(n)}+\epsilon_k^{(n+1)}\right\}} \\ &= \frac{\partial^2 A_i^{(n)}}{\partial\rho^2} + \frac{2}{\rho}\frac{\partial A_i^{(n)}}{\partial\rho} + \left.\frac{\partial V}{\partial\phi_i}\right|_{\left\{A_1^{(n+2)},\, \dots,\, A_i^{(n)},\, A_{i1}^{(n+2)},\, \dots\right\}} = 0. \end{aligned} \tag{4.91}$$

So we see that in the case of the single field an oscillating solution implies an exact solution to the equations of motion. Let us now turn to the multi-field case. The other fields $\phi_{j\neq i}$ will converge without oscillating corrections. Therefore, the corrections to change to the derivative of the tree-level potential will decay and thus will converge similarly to the single-field case. In the unlikely event that multiple fields start to have oscillating corrections, convergence could become slow. However, since the equations are still coupled one can make a similar argument to the above to see that convergence is still likely.

The other issue that needs to be checked is what happens when the mass matrix becomes zero. In this case the solution to the differential equations for the perturbations are

$$\frac{\partial^2\epsilon_i}{\partial\rho^2} + \frac{2}{\rho}\frac{\partial\epsilon_i}{\partial\rho} = B_i(\rho).$$

This is easily solved as

$$\epsilon_i = \beta_0 + \frac{\beta_{-1}}{\rho} + \int_0^{\rho}\frac{dy}{y^2}\int_0^{y} dx\, x^2 B(x),$$

where β_1 and β_{-0} will both be zero when the mass matrix is zero for all space–time points. Alternatively, if the mass matrix is singular one needs to take the singular value composition (or alternatively one can use a particular arrangement of the step functions used to approximate the mass matrix such that it is not singular in any region). All these considerations are of course avoided if one solves the differential equations governing the perturbations numerically.

4.7.2 A numerical example

The semi-analytic approach has a lot of steps, some of which are analytically intense. It is useful to demonstrate how it works using an example. Let us consider the two-field potential which was described in [31],

$$V(x, y) = (x^2 + y^2)[1.8(x - 1)^2 + 0.2(y - 1)^2 - \delta].$$

In the above, when the parameter $\delta = 0.4$ the path deviates a long way from the straight line ansatz so this is a sturdy test of the perturbative approach. The first step is to create the ansatz which is a straight line trajectory in field space. We therefore make the rotation in field space $(x, y) \to (u, \mathrm{v})$, where the global minimum is in the u direction. To reduce to a single-field problem we then simply set v to zero in the effective potential. We next want the global minimum to be at $u = 1$. This is achieved by a rescaling (and in general a shift) followed by division by $|E|$ which gives

$$\frac{V(u, 0)}{|E|} = 0.36u^2 - u^3 + 0.57u^4.$$

Our analytic formulas can now be directly used to make the ansatz. Doing so and then reversing the rescaling and rotation, the ansatz in the (x, y) basis is

$$x(\rho) = 1.046\left(1 - \tanh\left[\frac{\rho - 0.437}{1}\right]\right) \tag{4.92}$$

$$y(\rho) = 1.663\left(1 - \tanh\left[\frac{\rho - 0.437}{1}\right]\right). \tag{4.93}$$

Here we have the wall width being unity as a coincidence due to the rescaling. A useful trick to help convergence is to sanitize the above ansatz such that the derivative is zero when $\rho \to 0$. This is achieved by making a small modification to the ansatz:

$$\delta x(\rho) = \left.\frac{\partial x}{\partial \rho}\right|_{\rho=0} \exp\left[-\left.\frac{\partial x}{\partial \rho}\right|_{\rho=0} \rho\right] \tag{4.94}$$

$$\delta y(\rho) = \left.\frac{\partial y}{\partial \rho}\right|_{\rho=0} \exp\left[-\left.\frac{\partial y}{\partial \rho}\right|_{\rho=0} \rho\right]. \tag{4.95}$$

Each trajectory of $(x(\rho), y(\rho))$ is calculated for each successive perturbation in figure 4.11 and compared to the true solution. It takes three perturbations to converge to the real solution. In figure 4.12 the space–time dependence of the x and y fields is shown with each successive perturbation beginning with the single-field ansatz. Finally, figure 4.13 shows the convergence of the perturbation $B_x(\rho)$, plotted for successive perturbations of $x(\rho)$ starting with the ansatz. Incredibly, the magnitude of the error reduces by almost an order of magnitude with each perturbation.

4.7.3 Wall width and variation in β

In later chapters we will find that if nature has more than one Higgs field, it is possible to have a resonance in baryon number production. The peak of the resonance occurs when a particle has a degenerate mass with a virtual state.

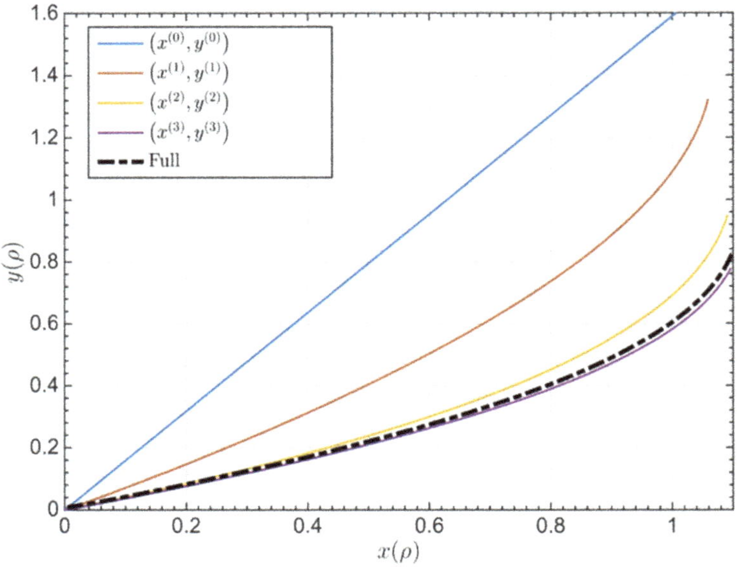

Figure 4.11. The bounce solution presented in field space for successive perturbations. The initial ansatz is the solid straight line while each perturbation is given by the solid curved lines. The numerical result from [31] is given by the dashed curve. Reproduced from [32].

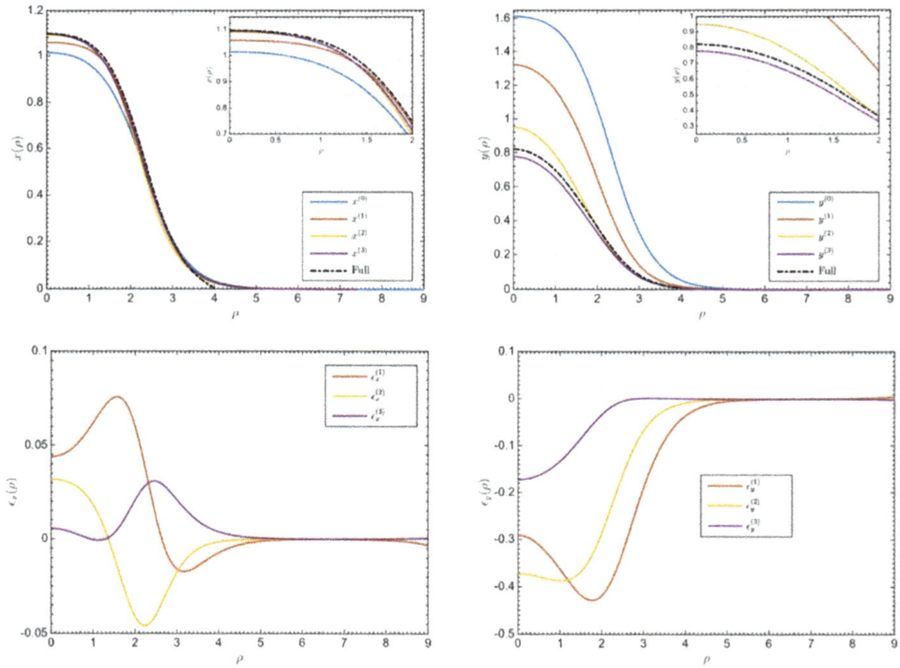

Figure 4.12. The single-field ansatz and the first three perturbations for x and y. The initial ansatzes for x and y are in the top left and right panels, respectively. The bottom left and right panels show the first three perturbations for x and y, respectively. Reproduced from [32].

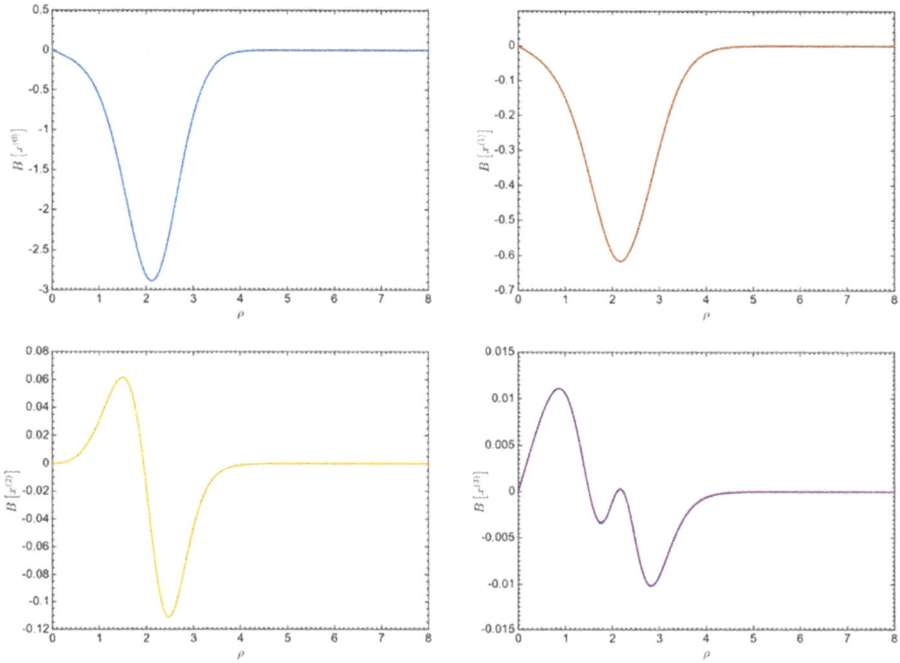

Figure 4.13. The error function B for the single-field ansatz for x and the first three perturbations. Reproduced from [32].

The strength of the resonance turns out to be weakly dependent on the wall width and linearly proportional to $\Delta\beta$, where $\tan \beta(T) = v_u(T)/v_d(T)$ and $\Delta\beta$ describes the maximum variation of β during the phase transition [35]. Generally, it needs to be calculated numerically, however, figure 4.14 shows a schematic of the tunneling path of the phase transition through two directions in field space. Generically, if one has more directions in field space it becomes easier for $\Delta\beta$ to be large.

4.8 Baryon washout condition

Let us expand on the criteria of a strongly first-order phase transition now that we have the sufficient tools. CP violating interactions with the bubble wall bias electroweak sphalerons such that baryons are created more frequently than anti-baryons, creating a net baryon asymmetry within the new phase. This baryon asymmetry will deplete within the new phase according to the rate equation

$$\frac{\mathrm{d}n_B}{\mathrm{d}t} = -k(T)n_B \tag{4.96}$$

for a temperature-dependent rate constant k, and n is the order of the phase transition. For a first-order phase transition n_B decays exponentially according to some half life $\ln 2/k(T)$, whereas a second-order phase transition decays inversely with time with a time-dependent half life. Naively at this point it might appear that a second-order phase transition is superior to a first-order phase transition when it

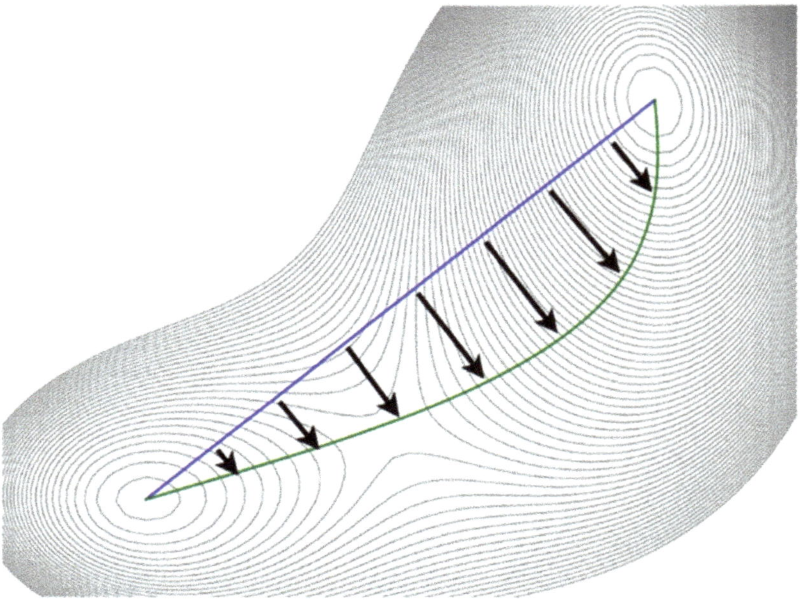

Figure 4.14. An example of how tan β can vary during a phase transition. The figure is a contour plot in field space of the effective potential of a two-scalar field system. The tunneling path from one minimum to another is a curved path in field space so tan β has a maximum and minimum value. The difference in the maximum and minimum value of β during the phase transition gives $\Delta\beta$. Reproduced from [31], copyright (2012) with permission from Elsevier.

comes to baryon number preservation. However, it is the difference in the half life between the two types of phase transitions that makes a strongly first-order phase transition necessary. The rate constant is just the sphaleron rate at high temperature which is given by some prefactor times an exponential of the strength of the phase transition

$$k(T) = -\frac{13n_f}{2VT^3}\frac{\omega_-}{2\pi}(\mathcal{N}\mathcal{V})_{\rm rot}(\mathcal{N}\mathcal{V})_{\rm tr}\left(\frac{v_C}{gT_C}\right)^3 e^{\frac{v_C(T_C)}{T_C}\frac{4\pi B\left(\frac{\lambda}{g}\right)}{g}}\kappa. \tag{4.97}$$

We can define a washout factor, S, as the ratio of the baryon asymmetry at the end of the phase transition, $n_B(t_f)$ compared to its value at the start, $n_B(t_i)$. The process of calculating the initial baryon asymmetry we will leave to the later chapters, but for now we can make some qualitative observations on the behavior of S as we modify v_C/T_C. A simple integration of the rate equation reveals a double exponential dependence on the strength of the phase transition

$$S \sim \exp e^{-\frac{v_C}{T_C}}. \tag{4.98}$$

So $S \to 1$ very quickly as we increase v_C/T_C. Suppose we have a model that can produce a baryon asymmetry e^X times the baryon asymmetry of the Universe at the time of the phase transition for a certain set of parameters. We then have a bound on the washout condition of

$$\frac{n_B(t_f)}{n_B(0)} > \mathrm{e}^{-X}. \tag{4.99}$$

Assuming that $k(T)$ is weakly dependent on time one can derive the bound

$$\frac{4\pi B}{g}\frac{v_C}{T_C} - 6\ln\frac{v_C}{T_C} > -\ln X - \ln\frac{t_f}{t_i} + \ln\frac{13n_f}{2}\frac{\omega_-}{2\pi}(\mathcal{NV})_{\rm rot}(\mathcal{NV})_{\rm tr} + \ln\kappa. \tag{4.100}$$

For $X \approx 1$ and $T_C \approx 100$ one finds that the above reduces to

$$\frac{v_C}{T_C} \gtrsim 1 \tag{4.101}$$

with theoretical uncertainties of about ~0.25 [17].

References

[1] Weinberg S 1974 Gauge and global symmetries at high temperature *Phys. Rev.* D **9** 3357

[2] Arnold P and McLerran L 1987 Sphalerons, small fluctuations, and Baryon-number violation in electroweak theory *Phys. Rev.* D **36** 2

[3] Shaposhnikov M E 1988 Structure of the high temperature gauge ground state and electroweak production of the baryon asymmetry *Nucl. Phys. B* **299** 4

[4] Arnold P and McLerran L 1988 The sphaleron strikes back: a response to objections to the sphaleron approximation *Phys. Rev.* D **37** 4

[5] Quiros M 1999 Finite temperature field theory and phase transitions arXiv:9901312v1(hep-ph)

[6] Das A 1997 *Finite Temperature Field Theory* (Singapore: World Scientific)

[7] Schwinger J 1961 Brownian motion of a quantum oscillator *J. Math. Phys.* **2** 407

[8] Mahanthappa K T 1962 Multiple production of photons in quantum electrodynamics *Phys. Rev.* **126** 329

[9] Bakshi P M and Mahanthappa K T 1963 Expectation value formalism in quantum field theory I *J. Math. Phys.* **4** 1

[10] Bakshi P M and Mahanthappa K T 1963 Expectation value formalism in quantum field theory II *J. Math. Phys.* **4** 12

[11] Keldysh L V 1965 Diagram technique for nonequilibrium processes *Sov. Phys.–JETP.* **20** 1018

[12] Craig R A 1968 Perturbation expansion for real-time Green's functions *J. Math. Phys.* **9** 605

[13] Chou K C, Su Z B, Hao B L and Yu L 1985 Equilibrium and nonequilibrium formalisms made unified *Phys. Rep.* **118** 1

[14] Millington P and Apostolos P 2013 Perturbative nonequilibrium thermal field theory *Phys. Rev.* D **88** 085009

[15] Peskin M E and Shroeder D V 1995 *An Introduction to Quantum Field Theory* (Boulder, CO: Westview)

[16] Dolan L and Jackiw R 1974 Symmetry behavior at finite temperature *Phys. Rev.* D **9** 3320

[17] Patel H H and Ramsey-Musolf M J 2011 Baryon washout, electroweak phase transition, and perturbation theory *J. High Energy Phys. JHEP07* **2011** 029

[18] Nielsen N K 1975 On the gauge dependence of spontaneous symmetry breaking in gauge theories *Nucl. Phys.* B **101** 1
[19] Fukuda R and Kugo T 1976 Gauge invariance in the effective action and potential *Phys. Rev.* D **13** 12
[20] Fendley P 1987 The effective potential and the coupling constant at high temperature *Phys. Lett.* B **196** 2
[21] Espinosa J R, Quirós M and Zwirne F 1992 On the phase transition in the scalar theory *Phys. Lett.* B **291** 115–24
[22] Carson L and McLerran L D 1990 Approximate computation of the small-fluctuation determinant around a sphaleron *Phys. Rev.* D **41** 647
[23] Carson L, Li X, McLerran L D and Wang R-T 1990 Exact computation of the small-fluctuation determinant around a sphaleron *Phys. Rev.* D **42** 2127
[24] Higgs P W 1964 Broken symmetries and the masses of gauge bosons *Phys. Rev. Lett.* **13** 16
[25] Coleman S 1977 Fate of the false vacuum: semiclassical theory *Phys. Rev.* D **15** 10
[26] Callan C G Jr and Coleman S 1977 Fate of the false vacuum. II. First quantum corrections *Phys. Rev.* D **16** 6
[27] Coleman S and De Luccia F 1980 Gravitational effects on and of vacuum decay *Phys. Rev.* D **21** 12
[28] McLerran L, Shaposhnikov M, Turok N and Voloshin M 1991 Why the baryon asymmetry of the Universe is ≈ 10–10 *Phys. Lett.* B **256** 477–83
[29] Dine M, Huet P and Singleton R Jr 1992 Baryogenesis at the electroweak scale *Nucl. Phys.* B **375** 625–48
[30] Profumo S, Ubaldi L and Wainwright C 2010 Singlet scalar dark matter: monochromatic gamma rays and metastable vacua *Phys. Rev.* D **82** 123514
[31] Wainwright C L 2012 CosmoTransitions: computing cosmological phase transition temperatures and bubble profiles with multiple fields *Comput. Phys. Commun.* **183** 2006–13
[32] Akula S, Balazs C and White G 2016 Semi-analytic techniques for calculating bubble wall profiles *Eur. Phys. J.* C **76** 12
[33] Dine M, Leigh R, Huet P, Linde A and Linde D 1992 Towards the theory of the electroweak phase transition *Phys. Rev.* D **46** 550
[34] John P 1999 Bubble wall profiles with more than one scalar field: a numerical approach *Phys. Lett.* B **452** 221–6
[35] Moreno J M, Quiros M and Seco M 1998 Bubbles in the supersymmetric standard model *Nucl. Phys.* B **526** 489–500

Chapter 5

Gravitational waves from a strongly first-order electroweak phase transition

The discovery of gravitational waves by LIGO [1] ushered in a new way of looking at the Universe. Since the Universe is transparent to gravitational waves right back to the beginning of time, any event that generated gravitational waves back then can in principle be observed today. This is a remarkable advantage when we compare rival techniques to probe the early Universe. The Universe stops being transparent to light if we try and peer before a million years after its birth. Indirectly, we can peer back to when the Universe was a minute old by counting the abundances of light elements produced during big bang nucleosynthesis. However, even this early time corresponds to a temperature of a mere MeV. Current terrestrial colliders probe energies many orders of magnitude higher.

It has been long known that a strong first-order electroweak phase transition would leave a gravitational wave background. Given this is a necessary condition for electroweak baryogenesis, this is in principle exciting. It remains unclear though whether a signal that can be observed with projected technology is consistent with any model of electroweak baryogenesis. The issue is that projected technology requires a very strong transition generating a lot of latent heat. Such a large amount of latent heat released implies a huge amount of pressure accelerating the expansion of the bubble wall. If the bubble wall becomes ultra-relativistic, it might be difficult for charge asymmetries to diffuse as they need to in order to bias the sphalerons in front of the bubble wall. Whether this is possible in such a powerful transition as we would need to observe it is an open question of research (for a recent exploration see reference [2]). In this chapter, we will assume it is a possible discovery channel for the paradigm to justify its discussion.

Let us begin this chapter with a very brief, albeit slightly formal introduction to gravitational waves in general before diving into the specific example of phase transitions. Einstein's equation predicting the relationship between any source and gravity has the form

doi:10.1088/978-0-7503-3571-3ch5

$$R_{\mu\nu} - \frac{1}{2}g_{\mu\nu}R = 8\pi\ T^{\mu\nu}. \tag{5.1}$$

The compact form of this equation is somewhat deceptive, we will need to manipulate this quite heavily in order to extract how a source corresponds to a gravitational wave spectrum. To begin with let us note that any gravitational wave sources we would realistically consider are small perturbations to a flat space–time metric

$$g_{\mu\nu} \sim \eta_{\mu\nu} + h_{\mu\nu}. \tag{5.2}$$

To first order in h, the linearized version of Einstein's equation has the form

$$\Box\bar{h}_{\mu\nu} + \eta_{\mu\nu}\partial^{\rho}\partial^{\sigma}\bar{h}_{\rho\sigma} - \partial^{\rho}\partial_{\nu}\bar{h}_{\mu\rho} - \partial^{\rho}\partial_{\mu}\bar{h}_{\nu\rho} = -16\pi G T^{\mu\nu}. \tag{5.3}$$

Here we have used the usual shorthand that $\bar{h}_{\mu\nu} = h_{\mu\nu} - \frac{1}{2}\eta_{\mu\nu}h$ and $h = \eta^{\mu\nu}h_{\mu\nu}$. At first glance, we have ten free parameters in a 4×4 symmetric tensor, however the degrees of freedom are a lot less than 10. First, general relativity is invariant to diffeomorphisms of the form $x^{\mu} \to x^{\mu} + \xi^{\mu}(x)$. This can be used to greatly reduce the number of free parameters. In particular we can choose a gauge where

$$\partial^{\nu}\bar{h}_{\mu\nu} = 0. \tag{5.4}$$

This allows us to eliminate four free parameters, leaving us with six and the linearized Einstein equation becomes

$$\Box\bar{h}_{\mu\nu} = -16\pi G T^{\mu\nu}. \tag{5.5}$$

For freely propagating gravitational waves away from their source, we can also ignore the stress energy tensor and we are free to make an additional coordinate transformation $x^{\mu} \to x^{\mu} + \xi^{\mu}(x)$, where $\Box\xi^{\mu} = 0$ without spoiling our gauge condition, to eliminate a further four free parameters. This leaves us with two free parameters left to match the number of degrees of freedom in a massless boson. We can use our freedom to make the trace vanish, $h = 0$ and transverse in the sense that the wave vector is perpendicular to the metric, $\partial^{i}h_{ij} = 0$. The neat feature of the transverse traceless gauge is that you can project onto the transverse traceless gauge with a projection operator

$$\Lambda_{ij,\,kl} = P_{ik}P_{jl} - \frac{1}{2}P_{ij}P_{kl} \tag{5.6}$$

where

$$P_{ij} = \delta_{ij} - n_i n_j \tag{5.7}$$

and n is a unit vector pointing in the same direction as the wave vector of the gravitational wave. It is straight forward to verify that this operator is indeed a projector, as $\Lambda_{ij,\,kl}\Lambda_{kl,\,mn} = \Lambda_{ij,\,mn}$ and it is transverse in its indices

$$n^{i}\Lambda_{ij,\,kl} = n^{j}\Lambda_{ij,\,kl} = \dots\ = 0. \tag{5.8}$$

Most importantly, the operator projects onto the transverse traceless gauge

$$h_{ij}^{TT} = \Lambda_{ij,\,kl} h_{kl}. \tag{5.9}$$

We can therefore use this projector to project out the part of the stress energy tensor that sources gravitational waves. Let us now proceed to derive the equation for the gravitational wave power spectrum. We will begin by writing everything as an expansion around the perturbation, h. For example, the Ricci tensor has the form[1]

$$R_{\mu\nu} = \overline{R}_{\mu\nu} + R^{1}_{\mu\nu} + R^{2}_{\mu\nu}, \tag{5.10}$$

where the subscripts indicate the power in h. We can then separate the equation into low frequency and high frequency modes. The zeroth-order term depends upon the large distance behavior of the metric only, so it is a low frequency term. The term that is linear order depends on gravitons, so is a short distance or high frequency effect. The second-order term, however, can contain high frequency and low frequency modes. For example, two large wave number modes can combine to give a small wave number mode. The total low frequency part of the equation then has the form

$$\overline{R}_{\mu\nu} = -\left[R^{2}_{\mu\nu}\right]^{\text{Low}} + 8\pi G\left(T_{\mu\nu} - \frac{1}{2}g_{\mu\nu}T\right)^{\text{Low}}. \tag{5.11}$$

Using our gauge condition, we can write the Ricci scalar contribution to the above equation as

$$R^{2}_{\mu\nu} = \frac{1}{2}\left(\frac{1}{2}\partial_\mu h_{\alpha\beta}\partial_\nu h^{\alpha\beta} + h^{\alpha\beta}\partial_\mu\partial_\nu h_{\alpha\beta}\right). \tag{5.12}$$

Taking the average and integrating by parts, we have

$$\left\langle R^{2}_{\mu\nu}\right\rangle = -\frac{1}{4}\left\langle \partial_\mu h_{\alpha\beta}\partial_\nu h^{\alpha\beta}\right\rangle. \tag{5.13}$$

Thus, we can write the expectation value of the low frequency part of the Einstein equation as

$$\left\langle \left(T_{\mu\nu} - \frac{1}{2}g_{\mu\nu}T\right)^{\text{Low}}\right\rangle = \frac{1}{32\pi G}\left\langle \partial_\mu h_{\alpha\beta}\partial_\nu h^{\alpha\beta}\right\rangle. \tag{5.14}$$

To derive the power spectrum, we need to project the LHS onto the transverse traceless mode. We then need to take the 00 component which corresponds to the energy part of the spectrum. After taking the Fourier transform one has

$$\frac{\mathrm{d}E}{\mathrm{d}\Phi} = \frac{G}{2\pi^2}\Lambda_{ij,\,kl}\int d\omega\; \omega^2 \overline{T}_{ij}(\omega, \omega\hat{n})\overline{T}_{kl}(-\omega, -\omega\hat{n}). \tag{5.15}$$

[1] For a more detailed discussion see reference [2].

Here $\overline{T}_{ij}$ is the Fourier transform of the stress energy tensor and Φ is the solid angle, usually denoted by the character Ω which we will reserve for an abundance. Indeed, it is both convenient and customary to consider the abundance of gravitational waves in each frequency bin observed today,

$$\Omega = \frac{k}{\rho_c}\frac{\mathrm{d}\rho_{GW}}{\mathrm{d}k}, \tag{5.16}$$

where

$$\frac{\mathrm{d}\rho_{GW}}{\mathrm{d}k\mathrm{d}\Phi} = 2G\, k^2 \Lambda_{ij,\, kl} \overline{T}_{ij} \overline{T}_{kl}. \tag{5.17}$$

The power spectrum above is written in terms of the wave number, which can easily be converted into a frequency when accounting for a factor of 2π and a redshift factor. We will discuss this conversion later in the chapter. We now have the formalism needed to derive the gravitational wave spectrum so long as we can identify the relevant parts of the stress energy tensor. During a phase transition, there are three components to the stress energy tensor that leads to gravitational waves. Two of these are easy to model,

1. The contribution from the scalar shell which can be identified as the scalar field contribution to the stress energy tensor $T_{\mu\nu}(x) = \partial_\mu \phi \partial_\nu \phi - g_{\mu\nu}\left[\frac{1}{2}\partial_\rho \phi \partial^\rho - V\right]$.
2. A contribution from soundwaves, where the plasma initially departs from its equilibrium state near the bubble wall then advances as sound shells that eventually collide. This contribution to the stress energy tensor has the form, $T_{\mu\nu}(x) = \sum \int \frac{\mathrm{d}^3 k}{(2\pi)^3 2E} k_\mu k_\nu f(k) = (e + p)U_\mu U_\nu + g_{\mu\nu} p$ where w is the enthalpy, U the fluid velocity and p is the pressure.

The other contribution to gravitational waves is from turbulence—when the sound shells collide larger scales break down into smaller scales in a reasonably long lived process. This last contribution is very difficult to model (for an attempt, see reference [3]). Of the other two components, the sound shells dominate unless there is an extreme amount of energy in the bubble wall, a rare scenario that can be achieved with a great deal of supercooling (see for example reference [4]). It is also believed, though not proven until numerical simulations give better information, that the sound shell contribution dominates over turbulence as well. We therefore focus on this contribution.

We will need to derive the fluid velocity profile that arises from an expanding bubble. We will take the approximation that the gravitational wave spectrum arising from many sound shell collisions can be modelled as an incoherent superposition of the contribution of individual sound shells that are in free propagation after the phase boundaries collide [5, 6]. To derive the fluid velocity we use the equations that arises from the conserved current involving the stress energy tensor. That is, the stress energy tensor obeys the usual conservation law that its divergence vanishes $T^{\mu\nu}_{;\mu} = 0$. We can separate the two components by including a friction term that captures the interaction between the fluid and the scalar shell. For simplicity we will work in Minkowski space, for this calculation performed in an expanding background see reference [7]. Since the expansion of the Universe is often quite slow compared to the time scale of the transition, ignoring the expansion of the Universe is often a reasonable approximation. The two components of the divergence of the stress energy tensor have the form,

$$T^{\mu\nu}_{;\mu|\text{field}} = (\partial^2\phi)\partial^\nu\phi - \frac{\partial V}{\partial\phi}\partial^\nu\phi = \delta^\nu. \tag{5.18}$$

$$T^{\mu\nu}_{;\mu|\text{fluid}} = \partial_\mu[(e+p)U^\mu U^\nu] + g^{\mu\nu}\partial_\mu p + \frac{\partial V}{\partial\phi} = -\delta^\nu. \tag{5.19}$$

It is trivial to see that the sums of the two terms are zero, as they should be, and the friction term can be modeled as the interaction between the fluid and the field gradients $\delta^\nu = \eta U^\mu \partial_\mu\phi\partial^\nu\phi$. Let us define a couple of shorthands for notational compactness

$$Z^i = \gamma^2(e+p)v^i, \quad E = e\gamma. \tag{5.20}$$

We will also take the fluid velocity to have the form, $U^\mu = \gamma(v, \hat{v}^i)$. We can write the equations of motion arising from the divergence of the stress energy tensor in spatial and temporal components as

$$-\ddot{\phi} + \nabla^2\phi - \frac{\partial V}{\partial\phi} = \eta\gamma(\dot{\phi} + \vec{v}\cdot\nabla\phi), \tag{5.21}$$

$$\dot{Z}^i + \nabla\cdot(\vec{v}\nabla Z^i) + \partial_i p + \frac{\partial V}{\partial\phi}\partial_i\phi = -\eta\gamma(\dot{\phi} + \vec{v}\cdot\nabla\phi)\partial_i\phi, \tag{5.22}$$

$$\dot{E} + p(\dot{\gamma} + \nabla\cdot\gamma\vec{v}) + \nabla\cdot(E\vec{v}) - \gamma\frac{\partial V}{\partial\phi}(\dot{\phi} + \vec{v}\cdot\nabla\phi) = \eta\gamma^2(\dot{\phi} + \vec{v}\cdot\nabla\phi)^2, \tag{5.23}$$

$$\begin{aligned} &\gamma^2(\dot{v} + \hat{v}\cdot v^2)(e+p) + v\dot{p} + \hat{v}\cdot\nabla p + \frac{\partial V}{\partial\phi}(v\dot{\phi} + \hat{v}\cdot\nabla\phi) \\ &\quad = -\eta\gamma(v\dot{\phi} + \hat{v}\cdot\phi)(\dot{\phi} + v\cdot\nabla\phi). \end{aligned} \tag{5.24}$$

Unless the bubble walls runaway such that the boost factor is very large, the field components of the above equation can be safely ignored and we have the significantly simpler system of equations

$$\dot{Z}^i + \nabla\cdot(\vec{v}Z^i) + \partial_i p = 0, \tag{5.25}$$

$$\dot{E} + p(\dot{\gamma} + \nabla\cdot\gamma\vec{v}) + \nabla\cdot(E\vec{v}) = 0, \tag{5.26}$$

$$\gamma^2(\dot{v} + \hat{v}\cdot v^2)(e+p) + v\dot{p} + \hat{v}\cdot\nabla p = 0. \tag{5.27}$$

We can make two further simplifications. First let us assume spherically symmetric sound shells, then we can note that because there is no characteristic scale in the problem, the solution should only depend on the ratio $\xi = r/t$. We can then denote $v(\xi)$ as the fluid velocity in the rest from of the center of the bubble and the gradients become

$$U_\mu \partial^\mu = \frac{\gamma}{t}(\xi - v)\partial_\xi. \tag{5.28}$$

We can then combine our equations into a single equation which has the form

$$2\frac{v}{\xi} = \gamma^2(1 - v\xi)\left(\frac{(\xi - v)^2}{c_s^2(1 - \xi v)^2} - 1\right)\partial_\xi v. \tag{5.29}$$

Here, c_s^2 is the speed of sound, given by $c_s^2 = \mathrm{d}p/\mathrm{d}e$,which we take to be its equilibrium value of 1/3 throughout. For a more precise treatment see reference [7]. To solve this equation, we require boundary condition at the bubble wall. We can use the bag equation of state where the pressure and energy are made continuous at the boundary by means of a bag constant

$$p_+ = \frac{1}{3}a_+T_+^4 - \epsilon, \quad p_- = \frac{1}{3}a_-T_-^4, \tag{5.30}$$

$$e_+ = a_+T_+^4 + \epsilon, \quad e_- = a_-T_-^4. \tag{5.31}$$

where ϵ and $\frac{1}{3}a_xT_x^4$ gives the Stephann Boltzmann pressure in each phase, that is

$$a = \frac{\pi^2}{30}\sum n_b + \frac{7}{8}n_f, \tag{5.32}$$

and $n_{b/f}$ are the number of relativistic bosonic/fermionic degrees of freedom in each phase, respectively. We can then match the stress energy tensor at the bubble wall boundary

$$w_+v_+^2\gamma_+^2 + p_+ = w_-v_-^2\gamma_-^2 + p_- \quad (T_+^{rr} = T_-^{rr}), \tag{5.33}$$

$$w_+v_+\gamma_+^2 = w_-v_-\gamma_-^2 \quad (T_+^{rt} = T_-^{rt}). \tag{5.34}$$

In the above we put the matching condition we were using in the parenthesis. From these two equations we can derive

$$v_+v_- = \frac{p_+ - p_-}{e_+ - e_-}, \quad \frac{v_+}{v_-} = \frac{p_+ + e_-}{e_+ + p_-}. \tag{5.35}$$

Using these boundary conditions, one can acquire a solution for the velocity field. It is not difficult to solve these equations numerically, however, we will later use a fit for the fluid velocity. Let us conclude by sketching the derivation of the gravitational wave power spectrum. We will parameterize the power spectrum as follows

$$\frac{d\rho_{GW}}{d\mathrm{Log}k} = \frac{1}{32\pi G}\frac{k^3}{2\pi^2}P_{\dot{h}}(k, t). \tag{5.36}$$

Assuming the fluid sheer stress, which we denote by $\tau^{ij} = (e + p)v^i v^j$, dominates the spatial components we can make an assumption of homogeneity and write the transverse traceless component of the two point unequal time correlator by parametrizing it as

$$\Lambda_{ij,\,kl}\left\langle \tau^{ij}_{k_1}(t_1)\tau^{kl}_{k_2}(t_2)\right\rangle = \Pi^2(k_1, t_1, t_2)(2\pi)^3\delta(k_1 + k_2). \tag{5.37}$$

Here the delta function ensures homogeneity which can be seen as follows: take the Fourier transform of both momenta variables and after the application of the delta function, it will only be a function of the relative coordinate. We will average over a period T for time scales much longer than the characteristic period of the gravitational waves,

$$P_{\dot{h}}(k, t) = (16\pi G)^2 \int_0^T dt_1 dt_2 \frac{\cos[k(t_1 - t_2)]}{2}\Pi^2(k, t_1, t_2). \tag{5.38}$$

Let us assume the unequal time correlator is only a function of the difference in times. We can further assume the fluid stress is characterized by a typical length scale, L_f, and does not evolve after creation. Under these assumptions we can parametrize

$$\Pi^2(k_2, t_1, t_2) = ((\bar{p} + \bar{e})\overline{U}_f^2)^2 L_f^3 \tilde{\Pi}^2(kL_f, k(t_1 - t_2)), \tag{5.39}$$

where $\tilde{\Pi}$ is a dimensionless function and the bars indicate averages, then

$$P_{GW}(kL_f) = \frac{1}{k \quad L_f}\int \mathrm{d}z \frac{\cos[z]}{2}\tilde{\Pi}^2(kL_f, z). \tag{5.40}$$

In the above $k(t_1 - t_2) = z$. Averaging over the lifetime, τ_v, and identifying the average energy density as the critical energy density $\bar{e} = \rho_c = 4H^2/8\pi G$, we can write the gravitational wave abundance as [5, 6, 8]

$$\Omega_{GW} = 3\left(1 + \frac{\bar{p}}{\bar{e}}\right)^2 \overline{U}_f^4 (H\tau_v)(HL_f)\frac{(kL_f)^3}{2\pi^2}P_{GW}(kL_f). \tag{5.41}$$

The factor $(H\tau_v)$ is often less than one and is a suppression factor from the finite lifetime of the sound wave. A careful derivation on an expanding background gives a slightly different factor (see reference [7]) which we will simply quote at the end. The function $P_{GW}(kL_f)$ needs to be numerically solved, as does the average fluid velocity. There are, however, some good fits for both of these in terms of thermal parameters in the phase transition. We will return to this shortly. First let us describe the spectrum in terms of thermal parameters. The first component to consider is the redshift factor. The total energy dilutes as radiation, whereas the critical energy density will only dilute as radiation until matter domination, after which it dilutes slower. This means the relationship between the gravitational wave abundance at the time compared to today is given by

$$h^2\Omega_{GW}(z = z_0, f) \leftrightarrow h^2\left(\frac{z_*}{z_0}\right)^4\left(\frac{H_*}{H_0}\right)^2 \Omega_{GW}(z = z_*, z_0 f/z_*). \tag{5.42}$$

Here and throughout, the starred subscript refers to the time of percolation which occurs approximately when the effective action of the bounce profile discussed in chapter 4, is two orders of magnitude larger than the temperature (a more exact approach is to calculate the volume fraction in each phase and take the temperature when there is an order one fraction in the low temperature phase). After applying this redshift factor and numerically solving for the power spectrum one finds the fit

$$h^2\Omega_{GW}(f) = 4.98 \times 10^{-5}\left(\frac{100}{g_*(T_*)}\right)^{1/3} \Gamma^2\overline{U}_f^4(H_*R_*)A\, S(f)\Upsilon(t_{SW}). \tag{5.43}$$

Here the numerical factor $A = 0.058$, $\Gamma^2 = 4/3$, the spectral form is

$$S(f) = \left(\frac{f}{f_{SW}}\right)^3\left[\frac{7}{4+3(f/f_{SW})^2}\right]^{7/2}, \tag{5.44}$$

which peaks at the frequency

$$f_{SW} = 2.65 \times 10^{-5}Hz\, \frac{1}{v_w}\left(\frac{g_s(T_*)}{100}\right)^{1/6}\left(\frac{T_*}{100\text{ GeV}}\right)\frac{z_p}{10}\frac{1}{H_*R_*}, \tag{5.45}$$

$z_p \sim 7$ is a simulation factor and the suppression factor from the finite time of the soundwaves takes the form

$$\Upsilon(t_{SW}) = 1 - \frac{1}{\sqrt{1 + 2\tau_{SW}H_*}}, \tag{5.46}$$

and of course H_* is the value of Hubble at the time of the phase transition. We now have several thermal parameters left to calculate. The bubble wall velocity is by far the most complicated and we will tackle it in chapter 8. To properly calculate the mean bubble separation, R_*, one should calculate the cubed route of the bubble number density at the time of percolation. From this one can estimate the lifetime of the soundwaves $\tau_{SW} = R_*/\overline{U}_f$. A more approximate approach that agrees to within around 20% of the true value is to take the derivative of the effective action of the bounce profile

$$R_* = (8\pi)^{1/3}\frac{v_w}{\beta}, \quad \frac{\beta}{H_*} = T\frac{\mathrm{d}(S_E/T)}{\mathrm{d}T}\bigg|_{T_*}. \tag{5.47}$$

The fluid velocity also can be approximated in terms of the trace anomaly,

$$\overline{U}_f^2 = \frac{3}{4}\kappa\alpha, \tag{5.48}$$

where the trace anomaly is

$$\alpha = \frac{\Delta V - Td\Delta V/dT}{\rho_c}, \tag{5.49}$$

and ΔV is the difference in the effective potential between the two phases, defined as to be positive semi-definite. Finally, the efficiency factor has the form [9]

$$\kappa = \begin{cases} \dfrac{\alpha}{0.73 + 0.083\sqrt{\alpha} + \alpha}, & v_w \sim 1 \\ \dfrac{v_w^{6/5} 6.9\alpha}{1.36 - 0.037\sqrt{\alpha} + \alpha}, & v_w < 0.1 \\ \dfrac{\alpha^{2/5}}{0.017 + (0.997 + \alpha)^{2/5}}, & v_w = c_s \end{cases}. \tag{5.50}$$

Of course, it is more accurate to numerically solve the fluid equations. Beyond this, numerical simulations seem to indicate that the sound shell model does a poor job capturing the suppression from energy that goes into vorticity modes. Unfortunately, this effect can only be captured numerically using a fit [10]. Beyond this, the power spectrum has two observables in its peak frequency and amplitude as the sound shell model predicts a spectral form that is independent of the thermal parameters. This mapping of two observables from four thermal parameters which are in turn derived from any number of multiparameter models can make the inverse problem look at first formidable. Simulations, however, seem to indicate a more promising picture, where the precise shape of the power spectrum is sensitive to all four thermal parameters, indicating that any observation of a gravitational wave signature would contain a lot of information about the underlying model that produced it [11]. Of course, this piece of good news is also bad news for phenomelogists who might wish for a simple mapping between their model and the predicted gravitational wave signal, simulations are formidably expensive! Still, this is a work in progress and the above spectral form should be taken with a grain of salt as a starting point in understanding the signals from a given theory (figure 5.1). An example of how the sound shell model can give a

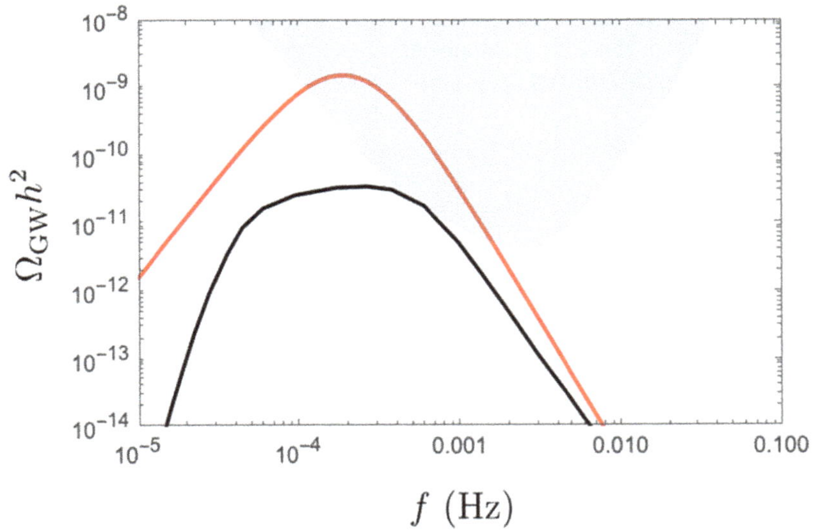

Figure 5.1. The gravitational wave spectrum for a benchmark as predicted from simulations (black) versus the spectrum predicted by the sound shell model (red) compared to the sensitivity of LISA (blue) [12]. The shape and amplitude differ dramatically for this benchmark which demonstrates the need for further development in predicting the gravitational wave spectrum. Benchmark and simulation data taken from reference [11].

different prediction of the spectrum compared to a simulation is given below. Note the existence of the breakpoint and the peak allows one to in principle reconstruct four thermal parameters.

References

[1] Abbott B PLigo Scientific and Virgo Collaborations *et al* 2016 Observation of gravitational waves from a binary black hole merger *Phys. Rev. Lett.* **116** 061102

[2] Maggiore M 2008 *Gravitational Waves* (Oxford: Oxford University Press)

[3] Caprini C, Durrer R and Servant G 2009 The stochastic gravitational wave background from turbulence and magnetic fields generated by a first order phase transition *JCAP* **12** 024

[4] Ellis J, Lewicki M, No J and Vaskonen V 2019 Gravitational wave energy budget in strongly supercooled phase transitions *JCAP* **06** 024

[5] HindMarsh M 2014 Gravitational waves from the sound of a first order phase transition *Phys. Rev. Lett.* **112** 041301

[6] HindMarsh M, Huber S and Rummukainen K 2015 Numerical simulations of acoustically generated gravitational waves at a first order phase transition *Phys. Rev.* D **92** 12

[7] Guo H, Sinha K, Vagie D and White G 2021 Phase transitions in an expanding Universe: stochastic gravitational waves in standard and non-standard histories *JCAP* **01** 001

[8] Ellis J, Lewicki M and No J 2019 On the maximal strength of a first order electroweak phase transition and its gravitational wave signal *JCAP* **04** 003

[9] Espinosa J, Konstandin T and Geraldine S 2010 Energy budget of cosmological first order phase transitions *JCAP* **06** 028

[10] Cutting D, HindMarsh M and Weir D 2019 Vorticity, kinetic energy, and suppressed gravitational wave production in strong first order phase transitions *Phys. Rev. Lett.* **125** 021302

[11] Gowling C and HindMarsh M 2021 Observational prospects for phase transitions at LISA: Fisher matrix analysis *JCAP* **10** 039

[12] Caprini C *et al* 2020 Detecting gravitational waves from cosmological phase transitions with LISA: an update *JCAP* **03** 024

Chapter 6

CP violation

The third Sakharov condition that needs to be satisfied is violation of C and CP. The Standard Model has CP violation through the CKM matrix—a CP violating phase in Higgs–quark Yukawa interactions cannot be completely rotated away through field redefinitions as one CP violating phase must remain [1]. The EDMs (electric dipole moments) produced by the CKM matrix are well below experimental lower bounds [2] and produce far too feeble a CP violation to produce the observed baryon asymmetry. This means there is plenty of experimental room for new sources of CP violation necessary for baryogenesis to hide in. The way of testing CP violation experimentally is through the search for permanent electric dipole moments in electrons, protons, and neutrons [3, 4]. This means that tests involve matter rather than anti-matter, so CPT invariance is assumed. Also, in general, the calculations involve zero temperature two-loop (or higher) calculations that involve the calculation of Wilson coefficients. The details of this can be found in many sources outside this book, so we only give details of the set-up of the problem and the big picture.

The summary of the logic is as follows. First, systematically list all diagrams that lead to CPV operators of the form given in tables 6.1 and 6.2 once heavy BSM (beyond the Standard Model) fields are integrated out. Examples of such diagrams for the minimal supersymmetric Standard Model (MSSM) are given in figure 6.1. This gives an effective operator for the electron EDM (electric dipole moment). For neutron and proton EDMs one needs to run from the weak scale to below the scale Λ_χ and calculate the appropriate Wilson coefficients that arise from systematically calculating appropriate matrix elements from the relevant terms in the effective Lagrangian.

The non-relativistic expression for an electric dipole moment for a spin-two particle is

$$H = -d\frac{\vec{E}\cdot\vec{s}}{|s|}, \tag{6.1}$$

doi:10.1088/978-0-7503-3571-3ch6

Table 6.1. List of dimension six CPV operators excluding four fermion interactions. List reproduced from [6].

Pure gauge		Gauge Higgs		Gauge Higgs fermion	
$\mathcal{O}_{\widetilde{G}}$	$f^{abc} G^a_{\mu\nu} \widetilde{G}^{b,\nu\beta} G^{c\mu}_{\beta}$	$\mathcal{O}_{\varphi\widetilde{G}}$	$\varphi^\dagger \varphi \widetilde{G}^a_{\mu\nu} G^{a,\mu\nu}$	$\mathcal{O}_{uG}$	$(\overline{Q}\sigma^{\mu\nu}\lambda^a u_R)\widetilde{\varphi} G^a_{\mu\nu}$
$\mathcal{O}_{\widetilde{W}}$	$\epsilon^{ijk} W^i_{\mu\nu} \widetilde{W}^{j,\nu\beta} W^{k\mu}_{\beta}$	$\mathcal{O}_{\varphi\widetilde{W}}$	$\varphi^\dagger \varphi \widetilde{W}^i_{\mu\nu} W^{i,\mu\nu}$	$\mathcal{O}_{fW}$	$(\overline{f}\ \sigma^{\mu\nu}\tau^i f_R)\Phi W^i_{\mu\nu}$
		$\mathcal{O}_{\varphi\widetilde{B}}$	$\varphi^\dagger \varphi \widetilde{B}_{\mu\nu} B^{\mu\nu}$	$\mathcal{O}_{dG}$	$(\overline{Q}\sigma^{\mu\nu}\lambda^a d_R)\widetilde{\varphi} G^a_{\mu\nu}$
		$\mathcal{O}_{\varphi\widetilde{W}B}$	$\varphi^\dagger \varphi \tau^i \widetilde{W}^i_{\mu\nu} B^{\mu\nu}$	$\mathcal{O}_{fB}$	$(\overline{f}\ \sigma^{\mu\nu} f_R)\Phi B_{\mu\nu}$

Table 6.2. List of dimension-six CPV four-fermion operators. List reproduced from [6].

Four fermion interactions	
$\mathcal{O}_{ledq}$	$\overline{L}^j e_R \overline{d}_R Q^j$
$\mathcal{O}_{quqd}$	$\overline{Q}^i u_R \epsilon_{ij} \overline{Q}^j d_R$
$\mathcal{O}^8_{quqd}$	$\overline{Q}^i \lambda^a u_R \epsilon_{ij} \overline{Q}^j \lambda^a d_R$
$\mathcal{O}_{lequ}$	$\overline{L} e_R \epsilon_{ij} \overline{Q} u_R$
$\mathcal{O}^3_{lequ}$	$\overline{L}\sigma_{\mu\nu} e_R \epsilon_{ij} \overline{Q}\sigma^{\mu\nu} u_R$

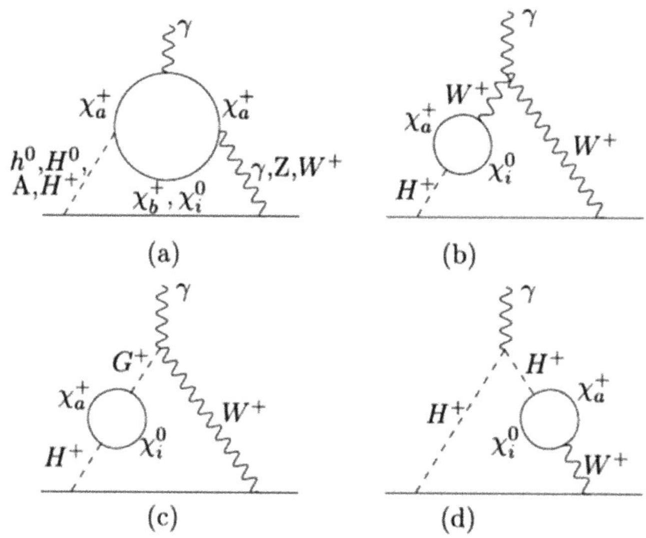

Figure 6.1. Some example operators that contribute to EDMs involving neutralino and charginos in the MSSM. Reproduced from [5], copyright (2008) by The American Physical Society.

which has the relativistic generalization [7]

$$\mathcal{L}_{\text{EDM}} = - d\frac{i}{2}\overline{\psi}\sigma^{\mu\nu}\gamma_5\psi F_{\mu\nu}. \tag{6.2}$$

One typically uses the framework of effective field theory to calculate contributions to EDMs [8, 9]. To calculate the observable EDM we need to consider the appropriate energy scale for a relevant experiment be it the QCD, nuclear, or atomic scale. One can then write out the set of higher dimensional operators that contribute to EDM, organizing them into powers of $1/\Lambda$ with the leading order expected to be dominant. At the nuclear scale the observables relevant to CP violation are the electron EDM d_e, the proton EDM d_p and the neutron EDM d_n, as well as electron–nucleon and nucleon–nucleon interactions with Lagrangians. These latter interactions are the parity violating and time reversal violating (PVTV) πNN interaction, the nuclear Schiff moment, and the PVTV electron–nucleus interaction.

For an energy scale above 1 GeV the Lagrangian organized into powers of $1/\Lambda$ is

$$\mathcal{L} = \mathcal{L}_{D=4} + \frac{1}{\Lambda}\mathcal{L}_{D=5} + \frac{1}{\Lambda^2}\mathcal{L}_{D=6} + O\left(\frac{1}{\Lambda^3}\right). \tag{6.3}$$

There is only one dimension-four operator which is topologically induced

$$\mathcal{L}_{D=4} = \frac{\alpha_s}{8\pi}\bar{\theta}G_{\mu\nu}\tilde{G}^{\mu\nu}. \tag{6.4}$$

This term produces far too much CP violation unless $\bar{\theta}$ is vanishingly small. This is known as the strong CP problem, with the most popular solution being the axion. We will assume that $\bar{\theta} = 0$ and ignore the dimension-four contribution. Naively the dimension-five terms are just $\mathcal{L}_{\text{EDM}}$, however for Standard Model particle content, gauge invariance forbids such a term without an additional Higgs field. The other dimension-six operators are listed in tables 6.1 and 6.2.

The gauge-Higgs–fermion interactions lead to terms of the form $\mathcal{L}_{\text{EDM}}$. After replacing the Higgs with its vacuum expectation value we obtain

$$\mathcal{O}_{qG} \to -i\sum_q \frac{g_3 \tilde{d}_q}{2}\bar{q}\sigma^{\mu\nu}\gamma_5\lambda^a q G^a_{\mu\nu} \tag{6.5}$$

$$\mathcal{O}_{fW} \to -i\sum_f \frac{d_f}{2}\bar{f}\,\sigma^{\mu\nu}\gamma_5\, f F_{\mu\nu} \tag{6.6}$$

with electric dipole moments

$$\tilde{d}_q = -\sqrt{2}\frac{v}{\Lambda^2}\text{Im}\left[C_{qG}\right] \tag{6.7}$$

$$d_f = -\sqrt{2}e\frac{v}{\Lambda^2}\,\mathrm{Im}[C_{f\gamma}] \tag{6.8}$$

where C_x are the Wilson coefficients.

Next one needs to relate the $D = 6$ CPV interactions to observables at the hadronic scale. The most useful framework one uses to calculate observables at this scale is heavy baryon chiral perturbation theory. Denoting the BSM scale as Λ and the chiral symmetry breaking scale as Λ_χ, one then produces an expansion in powers of Λ_χ/Λ. The four-fermion operators lead to effective interactions between electrons and nucleons given by

$$\begin{aligned}\mathcal{L}_{eN} &= \bar{e}i\gamma_5 e\overline{N}(C_S^0 + iC_S^1\tau_3)N + \bar{e}e\overline{N}i\gamma_5(C_P^0 + iC_P^1\tau_3)N \\ &\quad + \epsilon_{\mu\nu\alpha\beta}\bar{e}\sigma^{\mu\nu}e\overline{N}\sigma^{\alpha\beta}(C_T^0 + iC_T^1\tau_3)N,\end{aligned} \tag{6.9}$$

whereas the PVTV interactions involving pions and nucleons responsible for the nuclear Schiff moment are

$$\mathcal{L}_{\pi NN} = g_{\pi NN}^0\overline{N}\tau^a N\pi^a + g\pi NN^1\overline{N}N\pi^0 + g_{\pi NN}^2\left(\overline{N}\tau^a N\pi^a - 3\overline{N}\tau^3 N\pi^0\right). \tag{6.10}$$

Finally, the Lagrangian for the nucleon EDM at the hadronic scale defined in the nucleon rest frame is given by

$$\mathcal{L}_N = -\,2\overline{N}(d_0 + d_1\tau^3)\sigma^i NF^{0i}. \tag{6.11}$$

The hadronic couplings $d_p = d_0 + d_1$, $d_n = d_0 - d_1$, C_S, C_T, C_P, and $g_{\pi NN}^i$ are computed by calculating the matrix elements of the operators given in tables 6.1 and 6.2. If we ignore θ^- which arises from the four-dimensional CPV operator, the hadronic couplings have the form

$$\begin{aligned}d_{n,\,p} &= \frac{v^2}{\Lambda^2}\sum\beta_{n,\,p}\mathrm{Im}[C_{n,\,p}] \\ g_{\pi NN}^i &= \frac{v^2}{\Lambda^2}\sum\gamma\mathrm{Im}[C_{piNN}] \\ C_x &= \frac{v^2}{\Lambda^2}\sum\delta\mathrm{Im}[C_C].\end{aligned} \tag{6.12}$$

Experimental probes of EDMs look at paramagnetic atoms and polar molecules, which are most sensitive to d_e and C_S [7, 10, 11], diamagnetic atoms which are most sensitive to C_T and $g_{\pi NN}^{0,\,1}$, and neutron EDMs which are most sensitive to d_n and $g_{\pi NN}$ [12]. To conclude we give the current experimental bounds on the neutron and electron EDMs compared to the Standard Model value [2–4]

$$\begin{aligned}d_n^{\mathrm{SM}} &= 10^{-31}e\mathrm{cm} \quad d_n^{\exp} < 3.0\times 10^{26}e\mathrm{cm} \\ d_e^{\mathrm{SM}} &= 10^{-38}e\mathrm{cm} \quad d_e^{\exp} < 1.1\times 10^{-29}e\mathrm{cm},\end{aligned} \tag{6.13}$$

whereas the bounds on c_X and $g_{\pi NN}$ are

$$\begin{aligned} C_S &< 4.5 \times 10^{-7} \\ C_T &< 2 \times 10^{-6} \\ g^0_{\pi NN} &< 8 \times 10^{-9} \\ g^1_{\pi NN} &< 1 \times 10^{-9}. \end{aligned} \tag{6.14}$$

References

[1] Peskin M E and Shroeder D V 1995 *An Introduction to Quantum Field Theory* (Boulder, CO: Westview)

[2] Dar S 2000 The neutron EDM in the SM: a review arXiv:hep-th-0008248

[3] Pendlebury J M *et al* 2015 Revised experimental upper limit on the electric dipole moment of the neutron *Phys. Rev.* D **92** 092003

[4] ACME Collaboration 2018 Improved limit on the electric dipole moment of the electron *Nature* **562** 7727

[5] Li Y, Profumo S and Ramsey-Musolf M 2008 Higgs-Higgsino-gaugino induced two loop electric dipole moments *Phys. Rev.* D **78** 7

[6] Engel J, Ramsey-Musolf M J and Van Kolck U 2013 Electric dipole moments of nucleons, nuclei, and atoms: the Standard Model and beyond *Prog. Part. Nucl. Phys.* **71** 21–74

[7] Pospelov M and Ritz A 2005 Electric dipole moments as probes of new physics *Ann. Phys.* **318** 1

[8] Weinberg S 1976 Gauge theory of CP nonconservation *Phys. Rev. Lett.* **37** 11

[9] Barr S M and Zee A 1990 Electric dipole moment of the electron and of the neutron *Phys. Rev. Lett.* **65** 1

[10] Dzuba V A, Flambaum V V and Harabati C 2011 Relations between matrix elements of different weak interactions and interpretation of the parity-nonconserving and electron electric-dipole-moment measurements in atoms and molecules *Phys. Rev.* A **84** 5

[11] Jung M 2013 A robust limit for the electric dipole moment of the electron *J. High Energy Phys.* **JHEP05(2013)** 168

[12] Chupp T and Ramsey-Musolf M 2015 Electric dipole moments: a global analysis *Phys. Rev.* C **91** 3

Chapter 7

Particle dynamics during a phase transition, the vacuum expectation value insertion approach

Electroweak baryogenesis occurs during the electroweak phase transition, a process that happens when the Universe is at a temperature of $O(100)$ GeV in most models. Earlier we introduced the CTP formalism for handling the effective potential at finite temperature. However, we also need a way of describing the particle dynamics in a far-from-equilibrium setting [1]. In particular, conventionally equilibrium finite temperature quantum field theory (QFT) misses the substantial non-Markovian nature of such particle dynamics. In particular, these memory effects can cause a substantial boost to the baryon asymmetry through resonant effects [2, 3]. In this section our propagators become functions of space–time-dependent chemical potentials due to the modification of the Fermi and Bose distributions [4, 5]

$$n_{\mathrm{F,\,B}}(p_0) \to \frac{1}{\exp\left[\beta(p_0 + \mu)\right] \pm 1}. \tag{7.1}$$

We will show how the space–time-dependent chemical potentials govern the evolution of particle number currents. Eventually the chemical potentials themselves will be related to the number densities.

7.1 Particle current divergences and self-energy

Within the CTP formalism, divergences or particle currents can be related to the particles' self-energies [6]. We work with bosons to start with. Recall that our choice of time contour results in four independent Green's functions rather than the usual two. These can be written in matrix form for convenience:

$$G(x, z) \equiv \begin{pmatrix} G^{t}(x, z) & -G^{<}(x, z) \\ G^{>}(x, z) & -G^{\bar{t}}(x, z) \end{pmatrix} \tag{7.2}$$

The Schwinger–Dyson equation is then a matrix of equations.

$$\begin{aligned} \widetilde{G}(x, y) &= \widetilde{G}^0(x, y) + \int d^4w \int d^4z \widetilde{G}^0(x, w)\widetilde{\Sigma}(w, z)\widetilde{G}(z, y) \\ \widetilde{G}(x, y) &= \widetilde{G}^0(x, z) + \int d^4w \int d^4z \widetilde{G}(x, w)\widetilde{\Sigma}(w, z)\widetilde{G}^0(z, y). \end{aligned} \tag{7.3}$$

Here $\Sigma(x,z)$ is a matrix of self-energies corresponding to each propagator. Let us act with the Klein–Gordon operator, $K(x) \equiv \Box_x + m^2$, on both Schwinger–Dyson equations, remembering that the Klein–Gordon operator acting on $\widetilde{G}^0$ gives a delta function. We will also assume, as a simplification and therefore an approximation, that the mass associated with the Klein–Gordon operator is not space–time-dependent during the phase transition. The result can be expressed in matrix components denoted by the subscript ab,

$$\begin{aligned} K(x)\widetilde{G}(x, y)_{ab} &= -i\delta^4(x - y) - i\int d^4z \widetilde{\Sigma}(x, z)_{ac}\widetilde{G}(z, y)_{cb} \\ K(y)\widetilde{G}(x, t)_{ab} &= -i\delta^4(x - y) - i\int d^4z \widetilde{G}(x, y)_{ac}\widetilde{\Sigma}(z, y)_{cb}. \end{aligned} \tag{7.4}$$

Let us subtract the two equations and take the limit

$$\lim_{x\to y}[K(x)\widetilde{G}(x, y)_{ab} - K(y)\widetilde{G}(x, y)_{ab}] \tag{7.5}$$

Consider the (1,1) term of the above limit. The delta functions cancel and the resulting integrand, I, is

$$\begin{aligned} I = \lim_{x\to y} &\left[\theta(x_0 - z_0)\Sigma^>(x, z) + \theta(z_0 - x_0)\Sigma(x, z)\right] \\ &\times \left[\theta(z_0 - y_0)G^>(z, y) + \theta(y_0 - z_0)G^<(z, y)\right] \\ &- \left[\theta(x_0 - z_0)G^>(x, z) + \theta(z_0 - x_0)G^<(x, z)\right] \\ &\times \left[\theta(z_0 - y_0)\Sigma^>(z, y) + \theta(y_0 - z_0)\Sigma^<(z, y)\right] \\ &- \Sigma^>(x, z)G^<(z, y) + G^>(x, z)\Sigma^<(z, y). \end{aligned} \tag{7.6}$$

Using the identities that $\theta(x_0) + \theta(-x_0) = 1$ and $\theta(x_0)^2 = \theta(x_0)$ we can simplify the integrand to first remove products of step functions and then write everything proportional to a single step function,

$$\begin{aligned} I &= [\theta(x_0 - z_0) - 1]\Sigma^>(x, z)G^<(z, x) + \theta(z_0 - x_0)\Sigma^<(x, z)G^<(z, x) \\ &\quad \times [1 - \theta(x_0 - z_0)]G^>(x, z)\Sigma^<(z, x) - \theta(z_0 - x_0)G^<(x, z)\Sigma^>(z, x) \\ &= -\theta(z_0 - x_0)\Sigma^>(x, z)G^<(z, x) + \theta(z_0 - x_0)\Sigma^<(x, z)G^<(z, x)\theta(z_0 - x_0) \\ &\quad G^>(x, z)\Sigma^<(z, x) - \theta(z_0 - x_0)G^<(x, z)\Sigma^>(z, x). \end{aligned} \tag{7.7}$$

A simple calculation reveals that all four matrix elements are the same, so we have reduced our matrix of equations down to a single equation

$$\lim_{x\to y}\Big[K(x)\widetilde{G}(x, y) - K(y)\widetilde{G}(x, y)\Big]$$
$$= \int_{-\infty}^{x_0} dz_0 \int d^3z\Big[\widetilde{\Sigma}^{>}(x, z)\widetilde{G}^{<}(z, x) - \widetilde{G}^{<}(x, z)\widetilde{\Sigma}^{>}(z, x) \tag{7.8}$$
$$- \widetilde{\Sigma}^{<}(x, z)\widetilde{G}^{>}(z, x) + \widetilde{G}^{>}(x, z)\widetilde{\Sigma}^{<}(z, x)\Big].$$

Let us now consider the left-hand side of the above equation. We are free to choose any element, so let us choose the $G^{<}$ element. We can then show the left-hand side to be the divergence of the number current density

$$\begin{aligned} i\lim_{x\to y}\big[\Box_x - \Box_y\big]G^{<}(x, y) &= i\lim_{x\to y} \partial^x_\mu(\partial^\mu_x - \partial^\mu_y)G^{<}(x, y) \\ &= \lim_{x\to y} \mathrm{i}\partial^x_\mu(\partial^\mu_x - \partial^\mu_y)\langle\phi^\dagger_-(y)\phi_+(x)\rangle \\ &= i\lim_{x\to y} \partial^x_\mu\Big[\Big\langle\phi^\dagger_-(y)\partial^\mu_x\phi_+(x)\Big\rangle - (\partial^\mu_y\phi^\dagger_-(y))\phi_+(x)\Big] \\ &= i\partial_\mu\Big\langle\phi^\dagger_-(x)\overleftrightarrow{\partial}^\mu\phi_+(x)\Big\rangle = \partial^\mu j_\mu(x). \end{aligned} \tag{7.9}$$

Putting the pieces together we have shown that the divergence of the current density is equal to a function of the self-energies

$$\partial_\mu j^\mu = \int_{-\infty}^{x_0} dz_0 \int d^3z\Big[\widetilde{\Sigma}^{>}(x, z)\widetilde{G}^{<}(z, x) - \widetilde{G}^{<}(x, z)\widetilde{\Sigma}^{>}(z, x) \tag{7.10}$$
$$- \widetilde{\Sigma}^{>}(x, z)\widetilde{G}^{<}(z, x) + \widetilde{G}^{>}(x, z)\widetilde{\Sigma}^{<}(z, x)\Big].$$

Note that the current divergences depend on an integral over the entire history of the system. This leads to the system being highly non-Markovian. Also note that the above has no time ordered or anti-time ordered propagators or functions, which are the propagators that appear in ordinary zero temperature QFT. Having the off-diagonal Green's functions be non-zero as you have in finite temperature is a necessary condition for the particle number current density to have a non-zero divergence. Repeating all the above steps for fermions except using the sum of the Schwinger–Dyson equations we obtain a similar equation

$$\partial_\mu j^\mu = -\int_{-\infty}^{x_0} dz_0 \int d^3z \;\mathrm{Tr}\Big[\widetilde{\Sigma}^{>}(x, z)\widetilde{S}^{<}(z, x) - \widetilde{S}^{<}(x, z)\widetilde{\Sigma}^{>}(z, x) \tag{7.11}$$
$$- \widetilde{\Sigma}^{>}(x, z)\widetilde{S}^{<}(z, x) + \widetilde{S}^{>}(x, z)\widetilde{\Sigma}^{<}(z, x)\Big].$$

To derive a network of coupled transport equations one then needs to systematically evaluate all self-energy contributions for each particle species.

7.2 Transport coefficients and sources

To systematically deal with each type of self-energy that contributes to the transport equations it is useful to categorize them. First and most important are interactions with the space–time varying vacuum. These are analogous to the zero temperature mass insertion diagrams and are often called vacuum expectation values (VEV) insertions. We make the so called 'VEV insertion' approximation where off-diagonal contributions to mass matrices are treated perturbatively. To calculate these diagrams we make the assumption that the mass basis is not changing rapidly during the phase transition. The strength of these interactions is controlled by the Yukawa/triscalar coupling which can carry a CP violating phase, in which case they become the dominant (tree-level) CP violating source. Additionally, the strength of these sources is determined by the distance between the mass of the external and virtual state and the distance of both with the critical temperature. When the masses are nearly degenerate then there is a resonant boost to the CP violating sources [2–4, 7]. This is a particularly important feature as it implies that CP violating phases can be much smaller if one is near a resonance.

The VEV insertion diagrams also contain a CP conserving term which is also resonantly enhanced in strength when the masses of the virtual and external states are near degeneracy. This resonant enhancement of the relaxation term only slightly undermines the boost to the BAU produced via the resonant enhancement of the CP violating source.

All CP conserving interactions produce relaxation terms that ensure the number densities of the asymmetries, $n-n^{-}$, all decay away from the bubble wall. Mathematically there is a qualitative difference between CP violating sources and relaxation terms. CP violating sources are functions of space–time coordinates independent of chemical potentials, while relaxation terms are proportional to the chemical potentials, which themselves are proportional to the number density. To see this consider the zeroth component of the particle number current. Sources have the form $\partial_t n(x) \sim f(x)$ and, if a source exists in the network of transport equations, the solution for n cannot be satisfied by $n(x) \equiv n(x) - n(x) = 0$. In contrast, relaxation terms have the form $\partial_t n(x) \sim -\Gamma n(x)$, which has the trivial solution of $n(x) = 0$ in the absence of sources[1].

Apart from the VEV insertion diagrams, the other set of diagrams we consider that can be calculated perturbatively are the one-loop self-energies that involve Yukawa and triscalar interactions. These include supergauge interactions which, when sufficiently fast, are responsible for an approximate equilibrium between particles and their superpartners during the phase transition.

In order to tame the number of transport equations and interactions, one needs to consider which interactions are important. For a particle density to significantly affect the dynamics of the combined left-handed number density ahead of the bubble

---

[1] Note that we sometimes refer to relaxation terms as CP conserving sources, although strictly speaking they are relaxation terms. This is convenient as it allows us to be agnostic about the nature of each term on the RHS of equation (7.11) until we calculate it explicitly.

wall, the particle must diffuse before the advancing wall catches up. For a diffusion constant *D,* for a particular particle species, the diffusion length is given by Dt while the distance the wall travels in the same time t is v_{wt}. Taking $v_W = 0.05$ and $D = 50/T$ one has $t_{\rm diff} = 10^4/T$. Any process that occurs at a rate faster than the diffusion time is too significant to ignore, whereas if a particle couples to other particles only through rates slower than the diffusion time it can be approximately ignored in an analysis. Moreover, for a charge asymmetry to be transferred into an asymmetry in the left-handed number density, which seeds the ultimate baryon production by biasing the sphaleron processes, the particle number reactions involved in such a transfer must be fast enough that they occur before the advancing wall overtakes these particles. Incidentally, the electroweak sphaleron processes occur on a much larger time scale than the diffusion time. This means we can solve the dynamics that lead to an asymmetry in the left-handed number density first and use the calculated density as an input to a single differential equation for the production of the baryon number asymmetry.

Each self-energy is a formidable calculation and is sufficiently different from the sort of calculation one might encounter at zero temperature to warrant a significant degree of detailed explanation.

7.2.1 A biscalar with VEV insertions

Let us begin with the VEV insertions for scalar particles such as that given in figure 7.1. Here we can write the self-energy as $-g(x,y)G(x,y)$, where $g(x,y)$ contains all the information about the interaction with the space–time varying VEVs. The source equation is

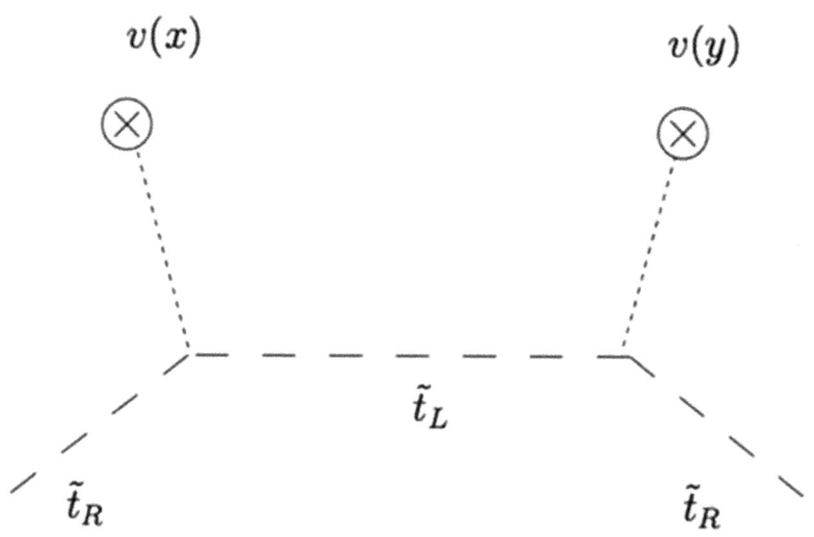

Figure 7.1. An example of a biscalar interaction with a space–time varying vacuum, also called a scalar VEV insertion. This particular diagram is relevant to any supersymmetric extension to the Standard Model. Reprinted with permission from [6], copyright (2005) by the American Physical Society.

$$\partial_\mu J^\mu = -\lim_{x \to z} \int d^3z \int_{-\infty}^{x_0} dz_0 \Big[g(x,z) G_1^>(x,z) G_2^<(z,x) - g(z,x) G_1^>(x,z)$$

$$G_2^<(z,x) + g(z,x) G_1^<(x,z) G_2^>(z,x) - g(x,z) G_1^<(x,z) G_2^>(z,x) \Big] \tag{7.12}$$

$$= -\int d^3z \int_{-\infty}^{x_0} dz_0 g(x,z) \Big[G_1^<(x,z) G_2^>(z,x) - G_1^>(x,z) G_2^<(z,x) \Big].$$

It is possible to separate our source equation into CP conserving and violating terms. Let us refer to the integrand as I. We can make use of the identities

$$\begin{aligned} I &= \frac{I}{2} + \frac{I^*}{2} + \frac{I}{2} - \frac{I^*}{2} \\ G(x,z) &= G^*(z,x) \end{aligned} \tag{7.13}$$

to show

$$S^{\mathrm{CP}} = -\lim_{x \to y}$$

$$\int d^3z \int_{-\infty}^{x_0} \{[g(x,z) + g(z,x)] Re[G_1^>(x,z) G_2^<(z,x) - G_1^<(x,z) G_2^>(z,x)] \tag{7.14}$$

$$+ i[g(x,z - g(z,x)] \mathrm{Im}[G_1^>(x,z) G_2^<(z,x) - G_1^<(x,z) G_2^>(z,x)]\} \equiv S^{CP} + S^{C/P}$$

We can see the need for CP violating phases, as without them the second term is zero. A less trivial result is that we also see that we need a minimum of two scalar fields (if we restrict ourselves to dimension four operators) to create a tree-level CP violating source through VEV insertion diagrams, since we need $g(x,z) \neq g(z,x)$, which generally requires interference between interactions with each VEV. These VEV insertions are relatively large sources of CP violation and thus give a relatively large baryon asymmetry. Now let us proceed to a full calculation of the VEV insertion sources beginning with a few definitions as well as a few identities to make the calculation more efficient. Recall that our Green's functions are given by

$$G^\lambda(x,z) = \int \frac{d^4k}{(2\pi)^4} e^{-ik\cdot(x-z)} g_B^\lambda(k_0, \mu_i) \rho(k_0, k), \tag{7.15}$$

where

$$\begin{aligned} \rho(k_0, k) &= \frac{i}{2\omega_k} \left[\frac{1}{k_0 - \mathcal{E}} + \frac{1}{k_0 + \mathcal{E}} - \frac{1}{k_0 - \mathcal{E}^*} - \frac{1}{k_0 + \mathcal{E}^*} \right] \\ g_B^>(k_0, \mu_i) &= 1 + n_B(k_0 - \mu_i) \\ g_B^<(k_0, \mu_i) &= n_B(k_0 - \mu_i) \\ n_B(x) &= \frac{1}{e^{\beta x} - 1}. \end{aligned} \tag{7.16}$$

In the above equations $\mathcal{E} = \omega + i\Gamma$ are the thermal widths and μ_i are the chemical potentials. We can then write the following useful identities

$$\begin{aligned} n_B(k_0 - \mu_i) &\approx n_B(k_0) - \mu_i \beta h_B(k_0) \\ g_B^{>}(-k_0, 0) &= - g_B^{<}(k_0, 0) \\ h(-k_0) &= h(k_0) \\ h_B(k_0) &= \frac{e^{\beta k_0}}{(e^{\beta k_0} - 1)^2}. \end{aligned} \tag{7.17}$$

Beginning with the CP conserving contribution to the source equation, let us expand SCP to first order in the chemical potentials, μ_i

$$S \approx S^0 + \delta S. \tag{7.18}$$

S^0 and δS both have poles from the function $\rho(k_0, k)$ at $\pm\mathcal{E}_i$ and $\pm\mathcal{E}_i^*$. We also have a tower of poles from the $n_B(k_0)$ and $h_B(k_0)$ functions at $k_0 = 2\pi n i/\beta$ for $n = 0, \pm 1, \pm 2, \ldots$. We will discuss these poles later. For now let us include them and rewrite our source as

$$S^{CP} \approx (S^0 - A_1) + (\delta S - A_2) + A_1 + A_2, \tag{7.19}$$

where A_i contains the contributions from the poles at $2\pi n i/\beta$. We will also show that S^0 does not contribute. Let us begin with the non-zero term, $\delta S - A_2$ and take the limit of $x = z$. We can immediately infer that we have a delta function in the spatial components. It is also convenient to perform a change in variable $z^0 - x^0 \rightarrow z^0$. Taking the integral over the delta function we have

$$\begin{aligned} \delta S - A_2 = &- 2g(x, x)\beta \, \mathrm{Re} \int_{-\infty}^{0} dz_0 \frac{d^4k}{(2\pi)^4} \frac{dq^0}{(2\pi)} e^{iz^0(k^0 - q^0)} \rho_1(k_0, k)\rho_2(q_0, k) \\ &\times \left[\mu_2 h_B(q_0) - \mu_1 h_B(k_0)\right]. \end{aligned} \tag{7.20}$$

It is easy to lose track of factors of i, 2π and (-1). This becomes even more difficult with the one-loop self-energies. It is therefore useful to write the following rules for taking these contour integrals:

- a factor of (-1) for each '*';
- an overall factor of (-1) from the requirement that the contour over k^0 is in the opposite direction as the contour over q_0; and
- a factor of $2\pi i^2/(2\omega_i)$ for each contour giving an overall factor of $(2\pi)^2/(4\omega_1\omega_2)$.

Using these rules to perform the contour integral the non-zero contribution to the source term is now

$$\delta S - A_2 = +\,2g(x,\,x)\beta\ \mathrm{Re}\int_{-\infty}^{0}\frac{\mathrm{d}z_0}{4\omega_1\omega_2}\frac{\mathrm{d}^3k}{(2\pi)^3}(-1)\mathrm{e}^{i(\mathcal{E}_1^*-\mathcal{E}_2)z^0}$$
$$\Big[\mu_2 h_B(\mathcal{E}_2) - \mu_1 h_B(\mathcal{E}_1^*)\Big] - \mathrm{e}^{i(-\mathcal{E}_1+\mathcal{E}_2^*)z^0}\Big[\mu_2 h_B(-\mathcal{E}_2) - \mu_1 h_B(-\mathcal{E}_1^*)\Big] \tag{7.21}$$
$$+\,\mathrm{e}^{i(-\mathcal{E}_1-\mathcal{E}_2)z^0}\Big[\mu_2 h_B(\mathcal{E}_2) - \mu_1 h_B(-\mathcal{E}_1)\Big] +$$
$$\mathrm{e}^{i(\mathcal{E}_1^*+\mathcal{E}_2^*)z^0}\Big[\mu_2 h_B(\mathcal{E}_2^*) - \mu_1 h_B(\mathcal{E}_1^*)\Big].$$

We obtain an additional factor of 4π when we integrate over the solid angle $\mathrm{d}\Omega$ and a factor of 2 from the real function. Finally, the integral over z_0 is now straightforward. A simple calculation gives

$$\delta S - A_2 = \frac{g(x,\,x)\beta}{2\pi^2}\ \mathrm{Im}\int_0^{\infty}\frac{k\mathrm{d}k}{\omega_1\omega_2}\Bigg[\frac{\mu_1 h_B(\mathcal{E}_1^*) - \mu_2 h_B(\mathcal{E}_2)}{\mathcal{E}_1^* - \mathcal{E}_2}$$
$$+\,\frac{\mu_1 h_B(\mathcal{E}_1) - \mu_2 h_B(\mathcal{E}_2)}{\mathcal{E}_1 + \mathcal{E}_2}\Bigg]. \tag{7.22}$$

Next we calculate $S^0 - A_1$. Since we only wish to show that the result is zero we can ignore all factors of proportionality

$$S^0 - A_1 \propto 2g(x,\,x)\beta\ \mathrm{Re}\int_{-\infty}^{0}\mathrm{d}z_0\frac{\mathrm{d}^4k}{(2\pi)^4}\frac{\mathrm{d}q^0}{(2\pi)}\mathrm{e}^{iz^0(k^0-q^0)}\rho_1(k_0,\,k)\rho_2(q_0,\,k)$$
$$\Big[g_B^>(k_0)g_B^<(q_0) - g_B^<(k_0)g_B^>(q_0)\Big]. \tag{7.23}$$

Once again, using our rules for taking these contour integrals which we defined before, we can perform the contour integrals over k^0 and q^0 and write the integrand of the above equation, I_A, as

$$\begin{aligned}
I_A &= (-1)\mathrm{e}^{i(\mathcal{E}_1-\mathcal{E}_2^*)z^0}[g^>(\mathcal{E}_1)g^<(\mathcal{E}_2^*) - g^<(\mathcal{E}_1)g^>(\mathcal{E}_2^*)]\\
&\quad+\ \mathrm{e}^{i(\mathcal{E}_1+\mathcal{E}_2)z^0}[g^>(\mathcal{E}_1)g^<(-\mathcal{E}_2) - g^<(\mathcal{E}_1)g^>(-\mathcal{E}_2)]\\
&\quad+\ \mathrm{e}^{i(\mathcal{E}_1-\mathcal{E}_2^*)z^0}[g^>(-\mathcal{E}_1^*)g^<(\mathcal{E}_2^*) - g^<(-\mathcal{E}_1^*)g^>(\mathcal{E}_2^*)]\\
&\quad+\ (-1)\mathrm{e}^{i(-\mathcal{E}_1^*+\mathcal{E}_2)z^0}[g^>(-\mathcal{E}_1^*)g^<(-\mathcal{E}_2) - g^<(-\mathcal{E}_1^*)g^>(-\mathcal{E}_2)]\\
&\quad\to \frac{-1}{i(\mathcal{E}_1 - \mathcal{E}_2^*)}[g^>(\mathcal{E}_1)g^<(\mathcal{E}_2^*) - g^<(\mathcal{E}_1)g^>(\mathcal{E}_2^*)]\\
&\quad+\ \frac{1}{i(\mathcal{E}_1^* - \mathcal{E}_2)}[g^>(\mathcal{E}_1^*)g^<(\mathcal{E}_2) - g^<(\mathcal{E}_1^*)g^>(\mathcal{E}_2)]\\
&= \mathrm{Re}\left[\frac{A + A^*}{i}\right] = 0.
\end{aligned} \tag{7.24}$$

So we have $S^0 = 0$. Finally, for the A_i terms we will note that they usually are ignored in the literature, however, we give a short comment that they disappear in the limit of vanishing thermal width. Let us write the Bose distribution functions and its derivative as Laurent series, which means factoring out a factor of $1/x$ and $1/x^2$, respectively, and Taylor expanding the remaining piece

$$\begin{aligned} \frac{1}{e^{\beta x}-1} &= \frac{1}{x}\frac{x}{e^{\beta x}-1} = \frac{1}{x}\sum_{n=0}^{\infty} a_n x^n \\ h_B(x) &= \frac{1}{x^2}\frac{x^2 e^{\beta x}}{(e^{\beta x}-1)^2} = \frac{1}{x^2}\sum_{n=0}^{\infty} b_n x^n. \end{aligned} \tag{7.25}$$

In both cases the sum is just a Taylor expansion with $a_0 = b_0 = 1/\beta$. For the $n \neq 0$ poles parallel to the real line, we can then just use the residue theorem and substitute the poles $p_0 = 2\pi i n\beta$ into the rest of the integrand and find that the function $\rho(k_0, k) \to \frac{i}{2\omega_k}[\delta(\omega - i2\pi n) + \delta(\omega + i2\pi n)]$ when you take the thermal widths to zero. These delta functions cannot be satisfied over the range of the integral so they do not contribute in this limit. For the $n = 0$ poles, these also do not contribute to the delta function as $\rho(k_0, k) = 0$ exactly when $k_0 = 0$ so this point does not contribute to the integral. Therefore, we can write the CP conserving contribution to the source equation as just the non-zero term we calculated earlier, which was linear in the chemical potentials. We now explicitly write the function $g(x,z)$ for the minimal supersymmetric Standard Model (MSSM) VEV insertion diagram for stops, as shown in figure 7.1

$$g(x, z) = y_t^2[A_t v_u(x) - \mu^* v_d(x)][A_t^* v_u(x) - \mu v_d(x)]. \tag{7.26}$$

It is useful to write the source equation as

$$S^{CP} = \Gamma_{\tilde{t}}^{+}(\mu_1 + \mu_2) + \Gamma_{\tilde{t}}^{-}(\mu_1 - \mu_2), \tag{7.27}$$

with

$$\begin{aligned} \Gamma_m^{\pm} = &- \frac{\beta N_C y_t^2}{4\pi^2} \mid A_t v_u(x) - \mu^* v_d(x)|^2 \int_0^{\infty} \frac{dk k^2}{\omega_L \omega_R} \\ &\times \mathrm{Im}\left\{\frac{h_B(\mathcal{E}_L) \mp h_B(\mathcal{E}_R^*)}{\mathcal{E}_L - \mathcal{E}_R^*} - \frac{h_B(\mathcal{E}_L) \mp h_B(\mathcal{E}_R)}{\mathcal{E}_L + \mathcal{E}_R}\right\}. \end{aligned} \tag{7.28}$$

Next let us calculate the CP violating source. In this case the difference between the space–time-dependent functions in the limit of $z \to x$ takes the form

$$\lim_{z\to x}[g(x, z) - g(z, x)] = 2iy_t^2 \mathrm{Im}(\mu A_t)[v_d(x)\partial_\lambda v_u(x) - \partial_\lambda v_d(x) v_u(x)](z - x)^\lambda. \tag{7.29}$$

Under the assumption of spatial isotropy only the time component of the above has a non-zero contribution, which is

$$\lim_{z\to x}[g(x, z) - g(z, x)] = 2iy_t^2 \mathrm{Im}(\mu A_t) v^2(x)\dot{\beta}(x)(z - x)^0. \tag{7.30}$$

Once again the spatial integral gives a delta function in the three momenta. The $(z - x)^0$ factor is handled by integrating the time component by parts. Performing the contour integrals over k_0 and p_0, and once again expanding $n_B(k_0, \mu_i)$ in powers of μ_i, we find that the zeroth order but not the first order is non-zero. After some algebra one obtains

$$S_{\tilde{t}_R}^{C/P} = \frac{N_C y_t^2}{2\pi^2}\mathrm{Im}[\mu A_t]v^2(x)\dot{\beta}\int_0^\infty \frac{dk k^2}{\omega_R\omega_L}\mathrm{Im}\left\{\frac{n_B(\mathcal{E}_R^*) - n_B(\mathcal{E}_L)}{(\mathcal{E}_L - \mathcal{E}_R^*)^2} + \frac{1 + n_B(\mathcal{E}_R) + n_B(\mathcal{E}_L)}{(\mathcal{E}_L + \mathcal{E}_R)^2}\right\}. \tag{7.31}$$

Both the CP conserving relaxation term and the CP violating source acquire a resonant boost in their size when the masses of the particles are nearly degenerate. Defining

$$S_0^{C/P} = \int_0^\infty \frac{\mathrm{d}kk^2}{\omega_R\omega_L}\,\mathrm{Im}\left\{\frac{n_B(\mathcal{E}_R^*) - n_B(\mathcal{E}_L)}{(\mathcal{E}_L - \mathcal{E}_R^*)^2} + \frac{1 + n_B(\mathcal{E}_R) + n_B(\mathcal{E}_L)}{(\mathcal{E}_L + \mathcal{E}_R)^2}\right\} \tag{7.32}$$

and

$$\Gamma_0 = -\int_0^\infty \frac{\mathrm{d}kk^2}{\omega_L\omega_R}\,\mathrm{Im}\left\{\frac{h_B(\mathcal{E}_L) + h_B(\mathcal{E}_R^*)}{\mathcal{E}_L - \mathcal{E}_R^*} - \frac{h_B(\mathcal{E}_L) + h_B(\mathcal{E}_R)}{\mathcal{E}_L + \mathcal{E}_R}\right\} \tag{7.33}$$

we plot in figure 7.2 the value of S_0 and Γ_0 as a function of m_R for $T = m_L = 100$. The effect of the resonance is dramatic, although there is some evidence that the center of the resonance may be severely damped when one goes beyond the approximations inherent in the above calculations [7]. This issue we return to in the discussion in chapter 16.

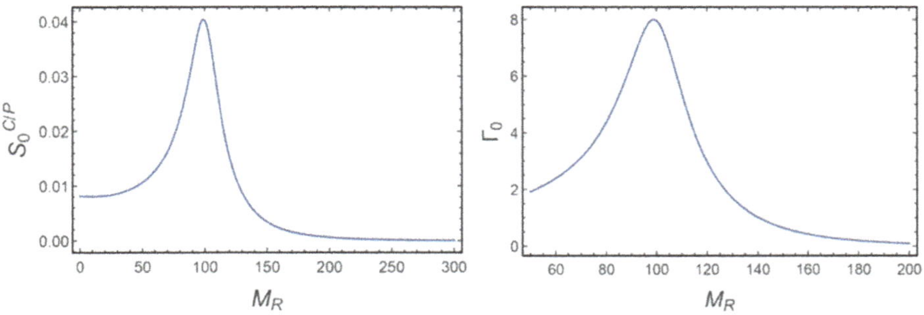

Figure 7.2. Plot of CPV source and CP conserving relaxation terms involving biscalar interaction with a space–time varying vacuum as a function of one of the masses. The above figures show the resonance for degenerate masses where $M_L = 100$, $T = 100$ and $\Gamma_L = \Gamma_R = 5$ with all units in GeV.

Before concluding this section we mention the scope of the approximation schemes we have used. If the mass of the virtual state is more than a factor of ~2 different to the mass of the in-state then the CP conserving relaxation term does not go to zero but actually becomes negative, which spuriously implies that baryon asymmetry can be generated even in the absence of a CP violating phase. To deal with this one needs to sanitize both the CP violating source and the CP conserving relaxation terms by setting them to zero by hand for a large mass splitting. The cutoff where this happens depends on the thermal widths. This is a generic feature of all VEV insertion diagrams, including the ones we will consider in the subsequent subsections.

7.2.2 Dirac fermions with VEV insertions

For an example of an interaction involving Dirac fermions contributing significantly to the baryon asymmetry, consider the Higgsino and gaugino interacting with the space–time varying vacuum in the MSSM. First write the Higgsino mass terms in the mass basis of the symmetric phase

$$L_{\tilde{H}} \ni |\mu| \left(\tilde{H}_d^0 \tilde{H}_u^0 - \tilde{H}_d^- \tilde{H}_u^+ + \tilde{H}_d^{0\dagger} \tilde{H}_u^{0\dagger} - \tilde{H}_d^{-\dagger} \tilde{H}_u^{+\dagger} \right). \tag{7.34}$$

Even though the Higgsino is a Majorana fermion, we can define three Dirac spinors out of the charged and neutral Higgsinos as well as a third out of the wino

$$\Psi_{\tilde{H}^+} = \begin{pmatrix} \tilde{H}_u^+ \\ \tilde{H}_d^{-\dagger} \end{pmatrix} \Psi_{\tilde{H}^0} = \begin{pmatrix} -\tilde{H}_u^0 \\ \tilde{H}_d^{0\dagger} \end{pmatrix} \Psi_{\tilde{W}^\pm} = \begin{pmatrix} \tilde{W}^+ \\ \tilde{W}^- \end{pmatrix}. \tag{7.35}$$

These fermions all obey the Dirac equation with masses ($|\mu|$, $|\mu|$, M_2), respectively. This means we can define a vector charge for these particle species. There are two gauginos that interact with the space–time varying vacuum that are generally ignored, the bino and the third wino, $\tilde{W}^3$. The justification for this is partly because one cannot define a vector charge for these particle species, which complicates matters, and partly because their contribution has been shown to be small [4]. The terms left in the Lagrangian that involve gaugino–Higgsino–vacuum interactions are

$$L_{\text{int}} \ni - g_2 \Psi \tilde{H} [v_d(x) P_L + v_u(x) e^{i\theta_\mu} P_R] \Psi_{\tilde{W}^+} \equiv - g_2 \Psi \tilde{H} [g_L(x) P_L + g_R(x) P_R] \Psi_{\tilde{W}^+}] \\ + h.\,c., \tag{7.36}$$

where in the second line we have implicitly defined $g_{L/R}$ for notational convenience. The self-energy for interactions of the form shown in figure 7.3 is

$$\Sigma_{\tilde{H}^\pm}(x, z) = -g_2^2 [g_L(x) P_L + g_R(x) P_R] S_{\tilde{W}^\pm}(x, z) [g_L(z)^* P_R + g_R(z)^* P_L]. \tag{7.37}$$

We then want to write the divergence of the current that arises from the above self-energy as a sum of a CP conserving relaxation term and a CP violating source term, just like we did before. However, this takes a little more care than before as the self-energy contains projection operators and the propagators contain Dirac matrices. Let us define

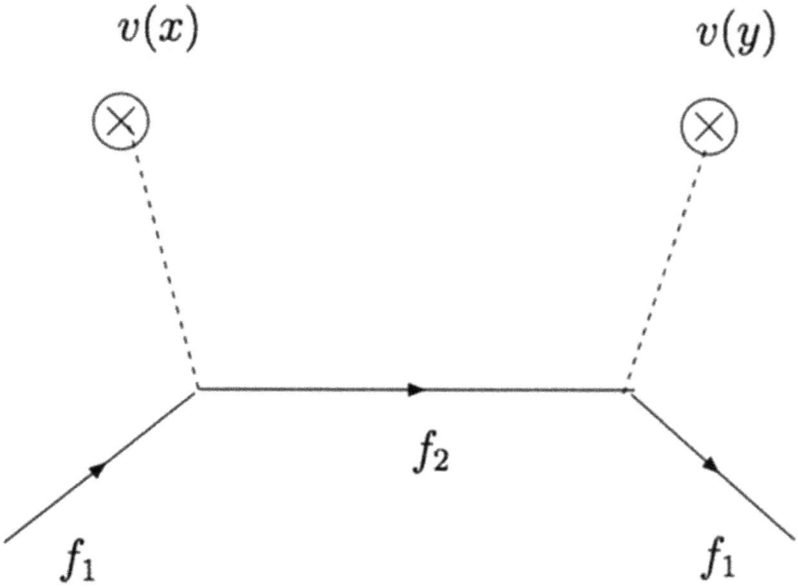

Figure 7.3. Depiction of a fermion interacting with a space–time varying vacuum via VEV insertion. Reprinted with permission from [6], copyright (2005) by the American Physical Society.

$$S^\lambda(k) = \overline{S}^\lambda(k)(\not{k} + m) \tag{7.38}$$

to allow us to separate our treatment of the matrices and the functions. When we calculate terms such as $\sim\int \Sigma^{>}_{\tilde{H}^\pm}(x, z) S^{<}_{\tilde{W}^\pm}(z, x) + \cdots$ we will encounter terms like the following (omitting measures and other details that detract from explaining the step):

$$\begin{aligned} &\int \mathrm{Tr}[g_L(x)P_L + g_R(x)P_R]\overline{S}_{\tilde{W}}(k)(\not{k} + |\mu|) \\ &\quad\times \left[g_L^*(z)P_R + g_R^*(z)P_L\right]\overline{S}_{\tilde{H}}(p)(\not{p} + M_2) \\ &= \int \mathrm{Tr}\Big\{\left[g_L(x)g_R^*(z)P_L + g_R(x)g_L(z)^*P_R\right]\not{k}\,\not{p} \\ &\quad + M_2|\mu|[g_L(x)g_R(z)^*P_L + g_R(x)g_L(z)^*P_R]\Big\} \\ &\quad\times \overline{S}^{>}_{\tilde{W}}(k)\overline{S}^{<}_{\tilde{H}}(p)e^{-i(x-z)\cdot(k-p)}. \end{aligned} \tag{7.39}$$

We can then define the space–time functions that multiply the kp terms and the mass terms

$$g_A(x, z) = \frac{g_2^2}{2}[g_L(x)g_L(z)^* + g_R(x)g_R(z)^*]$$

$$g_B(x, z) = \frac{g_2^2}{2}[g_L(x)g_R(z)^* + g_R(x)g_L(z)^*]$$

$$g_A(x, z) = \frac{g_2^2}{2}[g_L(x)g_L(z)^* + g_R(x)g_R(z)^*]$$
$$g_B(x, z) = \frac{g_2^2}{2}[g_L(x)g_R(z)^* + g_R(x)g_L(z)^*] \tag{7.40}$$

and note we have the identity that $g_j(x, z)^* = g(z, x)$ to write the current divergence for the gaugino–Higgsino–vacuum as a sum of a CP conserving relaxation term and a CP violating source term as before. It can be written in a compact form using the notation that $\mathrm{Tr}[\cdot]_A$ and $\mathrm{Tr}[\cdot]_B$ correspond to the terms involving the kp piece and the mass piece, respectively,

$$\begin{aligned} S_{\widetilde{H}^\pm}(x) &= \int d^3z \int_{-\infty}^{0} dz_0 \sum_{j\in(A,B)} \Big[g_j(x, z) + g_j(z, x)\Big] \\ &\quad \times \mathrm{ReTr}\Big[S^{>}_{\widetilde{W}^\pm}(x, z)S^{<}_{\widetilde{H}^\pm}(z, x) - S^{<}_{\widetilde{W}^\pm}(x, z)S^{>}_{\widetilde{H}^\pm}(z, x)\Big]_j \\ &= i\Big[g_j(x, z) - g_j(z, x)\Big]\mathrm{ImTr}\Big[S^{>}_{\widetilde{W}^\pm}(x, z)S^{<}_{\widetilde{H}^\pm}(z, x) \\ &\quad - S^{<}_{\widetilde{W}^\pm}(x, z)S^{>}_{\widetilde{H}^\pm}(z, x)\Big]_j. \end{aligned} \tag{7.41}$$

To solve these integrals, the steps are very similar to the biscalar–vacuum interaction so we just quote the result. The CP conserving term is given by

$$S^{CP}_{\widetilde{H}}(x) = \Gamma^{+}_{\widetilde{H}^\pm}(\mu_{\widetilde{W}^+} + \mu_{\widetilde{H}^+}) + \Gamma^{-}_{\widetilde{H}^\pm}(\mu_{\widetilde{W}^+} - \mu_{\widetilde{H}^+}) \tag{7.42}$$

with

$$\begin{aligned} \Gamma^{\pm}_{\widetilde{H}^\pm} &= \frac{1}{T}\frac{g_2^2}{2\pi^2}v(x)^2 \int_0^\infty \frac{dk k^2}{\omega_{\widetilde{H}}\omega_{\widetilde{W}}} \mathrm{Im}\Bigg\{\Big[\mathcal{E}_{\widetilde{W}}\mathcal{E}^*_{\widetilde{H}} \\ &\quad - k^2 M_2|\mu|\cos\theta_\mu \sin 2\beta\Big]\frac{h_F(\mathcal{E}_{\widetilde{W}}) \mp h_F(\mathcal{E}^*_{\widetilde{H}})}{\mathcal{E}_{\widetilde{W}} - \mathcal{E}^*_{\widetilde{H}}} \\ &\quad + \Big[\mathcal{E}_{\widetilde{W}}\mathcal{E}_{\widetilde{H}} + k^2 - M_2|\mu|\cos\theta_\mu \sin 2\beta\Big]\frac{h_F(\mathcal{E}_{\widetilde{W}}) \mp h_F(\mathcal{E}_{\widetilde{H}})}{\mathcal{E}_{\widetilde{W}} + \mathcal{E}_{\widetilde{H}}}\Bigg\} \end{aligned} \tag{7.43}$$

and the function $h_F(x) = \mathrm{e}^{x/T}/(\mathrm{e}^{x/T} + 1)^2$ defined analogously to the bosonic case. It is straightforward to derive the CP conserving term for the other Dirac fermions. If one follows the above derivation for $\widetilde{H}^0$ states, one will find that we merely make the replacement $g_2 \to g_2/\sqrt{2}$ when the virtual state is the wino. There is also a term for when the external state is the wino and the virtual state is the bino, which requires the frequency and the thermal widths to be changed appropriately (i.e. $\omega_{\widetilde{W}} \to \omega_{\widetilde{B}}$ and $\Gamma_{\widetilde{W}} \to \Gamma_{\widetilde{B}}$). This is similarly true for the CP violating sources,

$$S_{\widetilde{H}^{\pm}}^{C/P} = \frac{g_2^2}{\pi^2} v(x)^2 \dot{\beta}(x) M_2 |\mu| \sin\theta_\mu$$
$$\times \int_0^\infty \frac{dkk^2}{\omega_{\widetilde{H}} \omega_{\widetilde{W}}} \mathrm{Im}\left\{ \frac{n_F(\mathcal{E}_{\widetilde{W}}) - n_F(\mathcal{E}_{\widetilde{H}}^*)}{(\mathcal{E}_{\widetilde{W}} - \mathcal{E}_{\widetilde{H}^*})^2} + \frac{1 - n_F(\mathcal{E}_{\widetilde{W}}) - n_F(\mathcal{E}_{\widetilde{H}})}{(\mathcal{E}_{\widetilde{W}} + \mathcal{E}_{\widetilde{H}})^2} \right\}. \tag{7.44}$$

Just like the biscalar interactions, these terms have a resonant boost near where the mass is degenerate. Also, the approximation scheme used breaks down when the masses are very non-degenerate or exactly degenerate.

7.2.3 Chiral fermions with VEV insertions

The final VEV insertion term we will consider is the interaction between the top quarks and vacuum. There only arises a single CP conserving relaxation term. Interactions between the top quark and multi-particle states lead to additional poles of the propagator whose contribution is too large to ignore [8, 9]. For a massless top quark the top quark propagator in the presence of a thermal bath becomes

$$S^\lambda(x, y, \mu_i) = \int \frac{d^4k}{(2\pi)^4} e^{-ik\cdot(x-y)} g^\lambda(k_0, \mu)$$
$$\left[\frac{\gamma_0 - \gamma \cdot \hat{k}}{2} \rho_+(k_0, \vec{k}, \mu_i) + \frac{\gamma_0 + \gamma \cdot \hat{k}}{2} \rho_-(k_0, \vec{k}, \mu_i) \right] \tag{7.45}$$

where the spectral densities are related to each other via $\rho_-(k_0, \vec{k}, \mu_i) = [\rho_+(-k^{0*}, \vec{k}, -\mu_i)]^*$ and are defined by

$$\rho_+(k_0, \vec{k}, \mu_i) = i \left[\frac{Z_p(k, \mu_i)}{k_0 - \mathcal{E}_p(k, \mu_i)} - \frac{Z_p(k, \mu_i)^*}{k_0 - \mathcal{E}_p(k, \mu_i)^*} \right.$$
$$\left. + \frac{Z_h(k, -\mu_i)}{k_0 + \mathcal{E}_h(k, -\mu_i)^*} - \frac{Z_h(k, -\mu_i)}{k_0 + \mathcal{E}_h(k, -\mu_i)} + \cdots \right]. \tag{7.46}$$

Here the dots denote the non-pole contributions which we neglect. In the above equation, $Z_{p,\,h}$ are related to the poles of the equation

$$0 = k^0 - k - \frac{\alpha_s C_F \pi T^2}{2} \left[\left(1 - \frac{k^0}{k}\right) \log\left|\frac{k^0 + k}{k^0 - k}\right| + 2 \right]. \tag{7.47}$$

Explicitly, if we define the poles to the above equation by $k^0_{\mathrm{pole}} = \{E_p(k), -E_h(k)\}$, then we have

$$Z_{p,h} = \frac{E_{p,h}^2 - k^2}{m_f^2} \tag{7.48}$$

and the thermal mass (see section 7.4) is $m_f^2 = \alpha_s C_F \pi T^2/2$ where C_F is the Casimir of the group. There are two thermal widths $\Gamma_{p,h}$, which shift the poles in the imaginary direction as before,

$$\mathcal{E}_x = \sqrt{k^2 + Z_x(k, m)^2 m^2} - i\Gamma_x. \tag{7.49}$$

It is useful to write approximations for Z_p and Z_h from [8]. In the $k \ll m$ regime, we have the approximations

$$Z_p(k, m) \approx \frac{1}{2} + \frac{k}{6m} \tag{7.50}$$

$$Z_h(k, m) \approx \frac{1}{2} - \frac{k}{6m} \tag{7.51}$$

and the following in the regime $k \gg m$

$$Z_p(k, m) \approx 1 + \left(\frac{m^2}{2k^2}\right)\left(1 - \log\left[2\frac{k^2}{m^2}\right]\right) \tag{7.52}$$

$$Z_h(k, m) \approx \left(\frac{2k^2}{em^2}\right) e^{-2\frac{k^2}{m^2}}. \tag{7.53}$$

In the intermediate region we take a k/m dependent weighted average of the functions. Using interacting fermion propagators one can derive the rate

$$\begin{aligned}\Gamma^{\pm}_{m,\,t_R} = \frac{1}{T}\frac{N_C y_t^2 v^2(x)\sin^2\beta}{\pi^2}\int_0^\infty \mathrm{d}k k^2\, \mathrm{Im}\Bigg\{ &\frac{Z_p^R(k) Z_p^L(k)}{\mathcal{E}_p^R + \mathcal{E}_p^L}\left[h_F\left(\mathcal{E}_p^L\right) \mp h_F\left(\mathcal{E}_p^R\right)\right] \\ &+ \frac{Z_h^R(k)^* Z_p^L(k)}{\mathcal{E}_p^L - \mathcal{E}_h^{R*}}\left[h_F\left(\mathcal{E}_p^L\right) \mp h_F\left(\mathcal{E}_h^{R*}\right)\right] + (p \leftrightarrow h)\Bigg\}.\end{aligned} \tag{7.54}$$

7.2.4 Triscalar and Yukawa interactions

Let us now turn to the one-loop effects given in figure 7.4. Little new is learned from Yukawa and triscalar interactions so we will just look at the triscalar case and leave the Yukawa case as an exercise for the reader, merely quoting the result at the end. In the MSSM the most important such triscalar interaction—in the sense that it has the largest effect—is between the stops and the Higgs. There are three different diagrams with either the left- or right-handed stop or the Higgs as the 'in'-state. The three interactions will turn out to have a simple relation between them so we can just

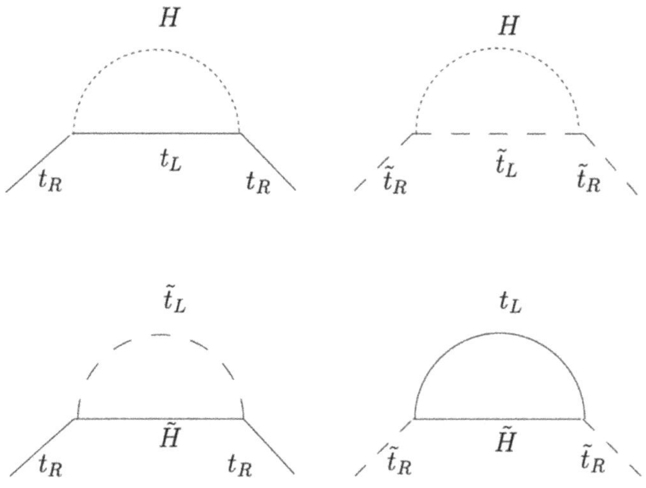

Figure 7.4. Most relevant one-loop MSSM Yukawa and triscalar interactions involving tops, stops, Higgs, and Higgsinos. Other triscalar and Yukawa interactions are suppressed by small Yukawa couplings. Reproduced with permission from [6], copyright (2005) by the American Physical Society..

consider one. Choosing the right-handed stop as the in-state, the self-energy can be written as just a product of propagators and the coupling

$$\Sigma^{\lambda} = |\lambda|^2 \, G_L^{\lambda} G_H^{\lambda}. \tag{7.55}$$

Here we have written the stop–Higgs triscalar coupling as λ to make the generalization clear. We will use the approximation of setting the thermal widths to be infinitesimal. Also in our analysis the following identity will be useful:

$$g^{>}(-\omega, \mu)|_{\mu=0} = -g^{<}(\omega, \mu)|_{\mu=0}, \tag{7.56}$$

which holds for both fermions and bosons. From here on, we will denote g^{λ} at zero μ simply as $g^{\lambda}(\omega)$. This identity is of course on top of the previous dictionary of identities introduced in the biscalar section. With this in mind, we can write to first order in μ that $\pm g^{<}_{B,\,F}(\omega, \mu) \approx g^{<}_{B,\,F}(\omega) - \beta\mu h_{B,\,F}(\omega)$, where $h_{B,\,F}(\omega) \equiv \exp(\beta\omega)/\{[\exp(\beta\omega) \mp 1]^2\}$. There is no CP violating source that arises from this self-energy, only a CP conserving relaxation term. Therefore, there cannot be a contribution from terms that are zeroth order in the chemical potentials, as such a term would cause a particle–anti-particle asymmetry even in the absence of CP violation. Thus, we can expand the propagators, G, and by implication the source S, to first order in the chemical potentials, μ_i. That is, $G \approx G^0 + \delta G$. We once again separate the terms that relate to the residues of the poles at $\mu_i + 2\pi n i/\beta = \{k_0, p_0, q_0\}$ and denote the sum of all these terms as 'A'. These terms we will ignore as they will not contribute. Our source to first order in the chemical potentials is now

$$S = (S^0 - A) + (\delta S - A) + 2A. \tag{7.57}$$

It can easily be shown mathematically that $S = \delta S$ as all other terms are zero. The proof of this, however, is very analogous to the calculation in the biscalar VEV insertion case, albeit messier and little more is learned from this. Since we have intuition as to why S^0 and A are zero we will omit this calculation. Using the identity that $\delta G^< = \delta G^>$ there are three terms for each product of the Green's functions leading to an expansion involving 12 terms in total

$$\delta S = |\lambda|^2 \int d^3z \int_{-\infty}^{X_0} dz_0[\delta G_L(x, z)G_H^>(x, z)G_R^<(z, x) + G_L^>(x, z)\delta G_H(x, z)G_R^<(z, x) + G_L^>(x, z)H^>(x, z)\delta G_R(z, x) + \cdots]. \tag{7.58}$$

We can factorize the expansion as the identity that $\delta G^>=\delta G^<$ introduces some degeneracies

$$\begin{aligned} \delta S &= |\lambda|^2 \int \mathrm{d}^3z \int_{-\infty}^{X_0} \mathrm{d}z_0 \delta G_L(x, z) \\ &\quad [G_H^>(x, z)G_R < (z, x) - G_H^<(x, z)G_R^>(z, x)] \\ &+ \delta G_L(z, x)[G_H^>(z, x)G_R < (x, z) - G_H^<(z, x)G_R^>(x, z)] \\ &+ \delta G_H(x, z)[G_L^>(x, z)G_R^<(z, x) - G_L^<(x, z)G_R^>(z, x)] \\ &+ \delta G_H(z, x)[G_L^>(z, x)G_R^<(x, z) - G_L^<(z, x)G_R^>(x, z)] \\ &+ \delta G_R(x, z)[G_L^>(x, z)G_H^>(z, x) - G_L^<(x, z)G_H^<(z, x)] \\ &+ \delta G_R(z, x)[G_L^>(z, x)G_H^>(x, z) - G_L^<(z, x)G_H^<(x, z)] \\ &\equiv \delta S_L + \delta S_H + \delta S_R. \end{aligned} \tag{7.59}$$

For now let us focus on δS_L. We can substitute in the expressions for the scalar propagators, $G^\lambda(x, z)$, and use the identity $G^\lambda(x, y) = G^{\lambda*}(z, x)$ to write

$$\delta S_L = 2|\lambda|^2 \beta\mu_L \mathrm{Re}\Big[\int d^3z \int_{-\infty}^{X_0} dz_0 \frac{d^4k}{(2\pi)^4}\frac{d^4p}{(2\pi)^4}\frac{d^4q}{(2\pi)^4} e^{-i(k+p-q)(x-z)} h(k_0) \times \rho(k_0, k)\rho(q_0, q)\rho(p_0, p)\big[g^>(p_0)g^<(q_0) - g^<(p_0)g^>(q_0)\big]\Big]. \tag{7.60}$$

Performing the integrals over d^3z makes a δ function for $k + p - q$. We then perform contour integrals for p_0 and q_0 omitting for now the overall sign of each term until we derive some rules for it. The contour over p_0 is carried out in the upper half of the Argand plane which contains the poles $\mathcal{E}_H^*$and $-\mathcal{E}_H$. Similarly the contour for q_0 is carried out over the lower half of the complex plane covering poles $\mathcal{E}_R$ and $-\mathcal{E}_R^*$. We also change variables to $\tau = z_0 - x_0$ and rename τ as z_0. We then have

$$\delta S_L = 2 \mid \lambda \mid^2 \beta\mu_L \mathrm{Re}\Bigg[\int_{-\infty}^{0} \mathrm{d}z_0 \frac{\mathrm{d}k_0}{(2\pi)} \frac{\mathrm{d}^3p}{(2\pi)^3} \frac{\mathrm{d}^3q}{(2\pi)^3} \frac{1}{4\omega_R\omega_L} h(k_0)\rho_L(k_0, q-p) \times$$
$$\Big\{(-1)^{L_1} \times \mathrm{e}^{-i(k_0+\mathcal{E}_H^*-\mathcal{E}_R)z_0}[g^>(\mathcal{E}_H^*)g^<(\mathcal{E}_R) - g^<(\mathcal{E}_H^*)g^>(\mathcal{E}_R)]$$
$$+ (-1)^{L_2} \times \mathrm{e}^{-i(k_0+\mathcal{E}_H^*+\mathcal{E}_R^*)z_0}[g^>(\mathcal{E}_H^*)g^<(-\mathcal{E}_R^*) - g^<(\mathcal{E}_H)^*)g^>(-\mathcal{E}_R^*)] \tag{7.61}$$
$$+ (-1)^{L_3} \times \mathrm{e}^{-i(k_0-\mathcal{E}_H-\mathcal{E}_R)z_0}[g^>(-\mathcal{E}_H)g^<(\mathcal{E}_R) - g^<(-\mathcal{E}_H)g^>(\mathcal{E}_R)]$$
$$+ (-1)^{L_1} \times \mathrm{e}^{-i(k_0-\mathcal{E}_H+\mathcal{E}_R^*)z_0}[g^>(-\mathcal{E}_H)g^<(-\mathcal{E}_R^*) - g^<(-\mathcal{E}_H)g^>(-\mathcal{E}_R^*)]\Big\}\Bigg].$$

To calculate the signs, L_i, we note that we acquire a factor of (-1) for each term from the contour integration itself as well as a factor of (-1) for every instance of a $\mathcal{E}^*$ term. Using these rules we can derive $L_1 = 2$, $L_2 = 3$, $L_3 = 1$, and $L_4 = 2$. Exploiting the presence of the Re[·] function we can turn the integral over z_0 into a δ function and use the identity $g^>(-x) = -g^<(x)$ to write

$$\begin{aligned}
\delta S_L \;=\; & 2\beta\mu_L \frac{\mid \lambda \mid^2}{2} \mathrm{Re}\Bigg[\int \frac{\mathrm{d}^3p}{(2\pi)^3} \frac{\mathrm{d}^3q}{(2\pi)^3} \frac{1}{4\omega_R\omega_H} \\
& \times \Big\{h(\mathcal{E}_R - \mathcal{E}_H^*)\rho_L(\mathcal{E}_R - \mathcal{E}_H^*, q-p)\big[g^>(\mathcal{E}_H^*)g^<(\mathcal{E}_R) - g^<(\mathcal{E}_H^*)g^>(\mathcal{E}_R)\big] \\
& - h(-\mathcal{E}_R^* - \mathcal{E}_H^*)\rho_L(-\mathcal{E}_R^* - \mathcal{E}_H^*, q-p) \\
& \times \big[-g^>(\mathcal{E}_H^*)g^>(\mathcal{E}_R^*) + g^<(\mathcal{E}_H^*)g^<(\mathcal{E}_R^*)\big] \\
& - h(\mathcal{E}_R + \mathcal{E}_H)\rho_L(\mathcal{E}_R + \mathcal{E}_H, q-p)\big[g^<(\mathcal{E}_H)g^<(\mathcal{E}_R) - g^>(\mathcal{E}_H)g^>(\mathcal{E}_R)\big] \\
& + h(-\mathcal{E}_R^* + \mathcal{E}_H)\rho_L(-\mathcal{E}_R^* + \mathcal{E}_H, q-p) \\
& \times \big[g^<(\mathcal{E}_H^*)g^>(\mathcal{E}_R^*) - g^>(\mathcal{E}_H)g^<(\mathcal{E}_R^*)\big]\Big\}\Bigg].
\end{aligned} \tag{7.62}$$

We now use the identities $\rho(x, y) = \rho^*(x, y)$, $\mathrm{Re}[A + A^*] = \mathrm{Re}[2a]$, $h(x) = h(-x)$ and $\rho(x,y) = -\rho(-x,y)$ which dramatically simplifies the form of the source

$$\begin{aligned}
\delta S_L = 2|\lambda|^2\beta\mu_L\, \mathrm{Re}\Bigg[\int & \frac{d^3pd^3q}{(2\pi)^6 4\omega_R\omega_H} \\
& h(\mathcal{E}_R - \mathcal{E}_H^*)\rho(\mathcal{E}_R - \mathcal{E}_H^*, q-p)\big[g^>(\mathcal{E}_H^*)g^<(\mathcal{E}_R) - g^<(\mathcal{E}_H^*)g^>(\mathcal{E}_R)\big] \\
& - h(\mathcal{E}_H + \mathcal{E}_R)\rho(\mathcal{E}_H + \mathcal{E}_R, q-p)\big[g^>(\mathcal{E}_H)g^>(\mathcal{E}_R) - g^<(\mathcal{E}_H)g^<(\mathcal{E}_R)\big]\Bigg].
\end{aligned} \tag{7.63}$$

The integrals over the angles can be performed by defining the angle between p and q such that $p \cdot q = pqx$. The rest of the angular integrals just produce a factor of $2(2\pi)^2$. We then perform a change of variables $\omega \equiv \omega_H + \sqrt{q^2 - p^2 - 2pqx + M_L^2}$ with new limits $\omega \pm \equiv \omega_H + \sqrt{q^2 - p^2 \pm 2pq + M_L^2}$ (chosen to absorb an extra minus sign) which allows us to write the more tractable integral

$$\delta S = |\lambda|^2 \beta\mu_L \operatorname{Re}\Bigg[\int_{\omega-}^{\omega^+} d\omega \frac{(pq)dpdq}{(2\pi)^4 2\omega_R\omega_H}$$
$$\times h(\mathcal{E}_R - \mathcal{E}_H^*)\rho'(\mathcal{E}_R - \mathcal{E}_H^*, q - p)\big[g^>(\mathcal{E}_H^*)g^<(\mathcal{E}_R) - g^<\big(\mathcal{E}_H^*\big)g^>(\mathcal{E}_R)\big] \tag{7.64}$$
$$- h(\mathcal{E}_H + \mathcal{E}_R)\rho'(\mathcal{E}_H + \mathcal{E}_R, q - p)\big[g^>(\mathcal{E}_H)g^>(\mathcal{E}_R) - g^<(\mathcal{E}_H)g^<(\mathcal{E}_R)\big]\Bigg],$$

where $\rho' \equiv \rho \times 2\omega$. Let us further simplify the above integral by performing manipulations on the following part of the integrand

$$\begin{aligned} &h(\mathcal{E}_R - \mathcal{E}_H^*)[g^>(\mathcal{E}_H^*)g^<(\mathcal{E}_R) - g^<(\mathcal{E}_H^*)g^>(\mathcal{E}_R)] \\ &= \frac{\exp[\beta(\mathcal{E}_R - \mathcal{E}_H^*)]}{\left\{\exp[\beta(\mathcal{E}_r - \mathcal{E}_H^*)] - 1\right\}^2}[n_B(\mathcal{E}_R) - n_B(\mathcal{E}_H^*)] \\ &= \frac{\exp[\beta(\mathcal{E}_R - \mathcal{E}_H^*)]}{\left\{\exp[\beta(\mathcal{E}_r - \mathcal{E}_H^*)] - 1\right\}^2}\left[\frac{\exp\beta\mathcal{E}_H^* - \exp\beta\mathcal{E}_R}{(\exp\beta\mathcal{E}_H^* 1)(\exp\beta\mathcal{E}_R - 1)}\right] \\ &= -\frac{\exp\beta\mathcal{E}_R}{(\exp[\beta(\mathcal{E}_R - \mathcal{E}_H^*)] - 1)(\exp\beta\mathcal{E}_R - 1)(\exp\beta\mathcal{E}_H^* - 1)} \equiv I_L. \end{aligned} \tag{7.65}$$

We can also derive the analogous terms I_H and I_R to be

$$\begin{aligned} I_H &= h(\mathcal{E}_H^*)[n_B(\mathcal{E}_R) - n_B(\mathcal{E}_H^*)] \\ &= -\frac{\exp\beta\mathcal{E}_R}{(\exp[\beta(\mathcal{E}_R - \mathcal{E}_H^*)] - 1)(\exp\beta\mathcal{E}_R - 1)(\exp\beta\mathcal{E}_H^* - 1)} \end{aligned} \tag{7.66}$$

and

$$\begin{aligned} I_R &= h(\mathcal{E}_R)[1 + n_B(\mathcal{E}_R - \mathcal{E}_H^*) + n_B(\mathcal{E}_H^*)] \\ &= +\frac{\exp\beta\mathcal{E}_R}{(\exp[\beta(\mathcal{E}_R - \mathcal{E}_H^*)] - 1)(\exp\beta\mathcal{E}_R - 1)(\exp\beta\mathcal{E}_H^* - 1)}, \end{aligned} \tag{7.67}$$

respectively. The other terms in the integrand are nothing more than the first term with $\mathcal{E}_H^* \to -\mathcal{E}_H$ and can be similarly simplified without putting pen to paper. Finally, we can make a change of variables from the respective momentum variables to frequencies. Recalling that $\delta S = S$, which we will show at the end of this calculation, our expression for the source term S_L is then [10]

$$S_L = (\mu_R - \mu_L - \mu_H)\Gamma_Y, \tag{7.68}$$

where

$$\Gamma_Y = |\lambda|^2 \beta \operatorname{Re}\Bigg[\int_{m_R, m_H}^{\infty} \frac{d\omega_R d\omega_H}{2(2\pi)^4} \int_{\omega-}^{\omega_+} d\omega$$
$$\big\{h(\mathcal{E}_R - \mathcal{E}_H^*)\rho_L{}'(\mathcal{E}_R - \mathcal{E}_H^*, \omega)[g^>(\mathcal{E}_H^*)g^<(\mathcal{E}_R) - g^<(\mathcal{E}_H^*)g^>(\mathcal{E}_R)] \tag{7.69}$$
$$- h(\mathcal{E}_H + \mathcal{E}_R)\rho_L{}'(\mathcal{E}_H + \mathcal{E}_R, \omega)\big[g^>(\mathcal{E}_H)g^>(\mathcal{E}_R) - g^<(\mathcal{E}_H)g^<(\mathcal{E}_R)\big]\big\}\Bigg].$$

One can easily obtain the form of S_H by making the following substitutions in the above analysis $k_0 \to p_0$, $q_0 \to k_0$ and $k_0 \to p_0$. We can immediately infer a remarkable identity $S_L = S_H$. To find the expression for S_L we make the changes $k_0 \to q_0$, $q_0 \to p_0$, and $p_0 \to k_0$. This takes a little more care as we now have to make both contour integrals over the top half of the complex plane. This means we acquire an additional minus sign and it is easy to see that $S_L = -S_R$. These sources are the same rate multiplied by the same linear combination of chemical potentials. The fact that the rate is the same corresponds to the fact that all three sources are the same physical process. Later we will see that these chemical potentials are proportional to their number densities for a given approximation. The linear combination of chemical potentials then serves to couple the differential equations governing the number densities of the three particle species. The overall sign then serves to ensure that the sign of this source term is always negative, meaning that it relaxes the combination of chemical potentials, which is what we expect since it is CP conserving. Let us next turn these relaxation terms into single integrals. To begin with let us note that our spectral density functions ρ consist of four Dirac delta functions:

$$\delta(\omega_L + \omega_R - \omega_H),\ \delta(\omega_L + \omega_R + \omega_H),\ \delta(\omega_L - \omega_R - \omega_H),\ \delta(\omega_L - \omega_R + \omega_H). \quad (7.70)$$

The second delta function is never non-zero within the limits of integration and can be ignored. We can use the first and fourth delta functions to perform the integral over ω_H. Then insisting that the limits of integration are real we obtain step functions

$$\Theta(m_2 - m_3 - m_1) + \Theta(m_3 - m_2 - m_1). \quad (7.71)$$

Performing the same procedure with the third delta function gives a step function

$$\Theta(m_1 - m_2 - m_3) \quad (7.72)$$

to ensure a real value for $\omega_\pm$ in each case. Our new limits of integration are

$$\omega_\pm = \frac{|\Delta|\,\omega_R \pm k\sqrt{\Delta^2 + m_R^2 m_L^2}}{m_R^2}, \quad (7.73)$$

with $\Delta = \frac{1}{2}(M_L^2 + M_R^2 - M_H^2)$. We then can write the triscalar decay rate as

$$\begin{aligned}\Gamma_y &= |\lambda|^2 \beta \int_{m_R}^{\infty} \mathrm{d}\omega_R \int_{\omega^-}^{\omega^+} d\omega \\ &\Big\{ n_B(\omega_R)\big[1 + n_B(\omega_L)\big] n_B(\omega_L - \omega_R)\big[\Theta(m_R - m_L - m_H) - \Theta(m_L - m_R m_H)\big] \\ &\quad - n_B(\omega_R) n_B(\omega_L)\big[1 + n_B(\omega_L + \omega_R)\big]\Theta(m_H - m_R - m_L)\Big\}.\end{aligned} \quad (7.74)$$

Finally, performing the integral over dω we can reduce our integral to be over one variable

$$\begin{aligned}\Gamma_Y = &-\frac{|\lambda|^2\beta}{16\pi^3}\int_{m_R}^{\infty} d\omega_R h_B(\omega_R)\times\left\{\log\left(\frac{e^{\omega_R/T}-e^{\omega^+/T}}{e^{\omega_R/T}-e^{\omega^-/T}}\frac{e^{\omega^-/T}-1}{e^{\omega^+/T}-1}\right)\right.\\ &\left[\Theta\left(m_1-m_2-m_\phi\right)-\Theta\left(m_2-m_1-m_\phi\right)\right]+\\ &\left.\log\left(\frac{e^{-\omega_R/T}-e^{\omega^+/T}}{e^{-\omega_R/T}-e^{\omega^-/T}}\frac{e^{\omega^-/T}-1}{e^{\omega^+/T}-1}\right)\Theta(m_\phi-m_1-m_2)\right\}\equiv I_B(m_L, m_R, m_H).\end{aligned} \tag{7.75}$$

Note that a triscalar decay cannot occur unless the parent species is larger than the combined mass of the daughter species, giving these integrals a clear physical interpretation. Next let us show the analogous function for Yukawa interactions, skipping the lengthy and very similar derivation

$$\begin{aligned}\Gamma_Y = &-\frac{|\lambda|^2\beta}{16\pi^3}(m_1^2+m_2^2-m_\phi^2)\int_{m_1}^{\infty} d\omega_1 h_F(\omega_1)\\ &\times\left\{\log\left(\frac{e^{\omega_\phi^-/T}+e^{\omega_1/T}}{e^{\omega_\phi^+/T}+e^{\omega_1^-/T}}\frac{e^{\omega_\phi^+/T}-1}{e^{\omega_\phi^-/T}-1}\right)\right.\\ &\left[\Theta(m_1-m_2-m_\phi)-\Theta(m_\phi-m_1-m_2)\right]\\ &\left.+\log\left(\frac{e^{\omega_\phi^-/T}+e^{-\omega_1/T}}{e^{\omega_\phi^+/T}-e^{-\omega_1/T}}\frac{e^{\omega_\phi^+/T}-1}{e^{\omega_\phi^-/T}-1}\right)\Theta(m_2-m_1-m_\phi)\right\}\\ \equiv\ & |\lambda|^2 I_F(m_1, m_2, m_\phi)\end{aligned} \tag{7.76}$$

with $\omega_\phi^\pm$ given by

$$\begin{aligned}\omega_\pm = \frac{1}{2m_1^2} &\left\{\omega_1\left|m_\phi^2+m_1^2-m_{\frac{1}{2}2}^2\right|\right.\\ &\left.\left[(\omega_1^2-m_1^2)(m_1^2-(m_2+m_\phi)^2)(m_1^2-(m_2-m_\phi)^2)\right]^{1/2}\right\}.\end{aligned} \tag{7.77}$$

For a supergauge interaction such as

$$l \ni \phi\overline{\Psi}(g_L P_L + g_R P_R)\widetilde{V}, \tag{7.78}$$

which is responsible for one-loop interactions such as the one shown in figure 7.6, the sources resulting from such interactions can be related to the fermion three-body rates [11]

$$S_{\widetilde{V}} = \left(|g_L|^2+|g_R|^2\right)I_F(m_\psi, m_\phi, m_{\widetilde{V}}) + 2\mathrm{Re}\left(g_L g_R^*\right)\widetilde{I}_F(m_\psi, m_\phi, m_{\widetilde{V}}), \tag{7.79}$$

where we have defined

$$\widetilde{I}_F(m_1, m_\phi, m_2) = \frac{2m_1m_2}{m_1^2 + m_2^2 - m_\phi^2} I_F(m_1, m_\phi, m_2). \tag{7.80}$$

To conclude this section we will include all relevant one-loop three-body interactions in the MSSM, excluding the gauge–boson interactions. By relevant we mean all interactions that involve the order 1 coupling constants $\{y_t, g_1, g_2, g_3\}$. Since the three-body rates scale as the coupling constant squared, even three-body rates involving y_b and y_τ are heavily suppressed. In order to catalog the set of three-body rates it is useful to introduce a couple of shorthands. First let us denote

$$I_{F,B}^{1,2,3} = I_{F,B}(m_1, m_2, m_3), \tag{7.81}$$

as well as denoting the charged Higgs mixing angle as follows:

$$\begin{pmatrix} H_u^+ \\ H_d^- \end{pmatrix} = \begin{pmatrix} c_\alpha & s_\alpha \\ -s_\alpha & c_\alpha \end{pmatrix} \begin{pmatrix} H_1^+ \\ H_2^+ \end{pmatrix}. \tag{7.82}$$

We will also denote the left-handed quark doublet by Q_L for reasons we will explain in section 7.4. The set of three-body supergauge interactions between Higgs and Higgsinos are

$$\begin{aligned} \left\{\Gamma_{\widetilde{V}}^{H_1, \widetilde{H}}, \Gamma_{\widetilde{V}}^{H_2 \widetilde{H}}\right\} = &\left\{g_1^2\left(I_F^{\widetilde{H}, H_1, \widetilde{H}} - s_\alpha \widetilde{I}_F^{\widetilde{H}, H_1, \widetilde{B}}\right)\right. \\ &+ 3g_2^2\left(I_F^{\widetilde{H}, H_1, \widetilde{W}} + s_\alpha \widetilde{I}_F^{\widetilde{H}, H_1, \widetilde{W}}\right), g_1^2\left(I_F^{\widetilde{H}, H_2, \widetilde{B}} + s_\alpha \widetilde{I}_F^{\widetilde{H}, H_2, \widetilde{B}}\right) \\ &\left. + 3g_2^2\left(I_F^{\widetilde{H}, H_2, \widetilde{W}} - s_\alpha \widetilde{I}_F^{\widetilde{H}, H_2, \widetilde{W}}\right)\right\}. \end{aligned} \tag{7.83}$$

The set of three-body supergauge interactions involving third-generation quarks are

$$\begin{aligned} &\left\{\Gamma_{\widetilde{V}}^{Q_L, \widetilde{Q}_L}, \Gamma_{\widetilde{V}}^{t_L, \widetilde{t}_R}\right\} = N_C \times \\ &\left\{\frac{g_1^2}{9} I_F^{Q_L, \widetilde{Q}_L, \widetilde{B}} + 3g_2^2 I_F^{Q_L, \widetilde{Q}_L, \widetilde{W}}\right. \\ &\left. + \frac{2(N_C^2 - 1)}{N_C} g_3^2 I_F^{Q_L, \widetilde{Q}_L, \widetilde{G}}, \frac{8g_1^2}{9} I_F^{t_R, \widetilde{t}_R, \widetilde{B}} + \frac{(N_c^2 - 1)}{N_C} g_3^2 I_F^{t_R, \widetilde{t}_R, \widetilde{G}}\right\}, \end{aligned} \tag{7.84}$$

and finally the three-body rates involving third-generation interactions between squarks, quarks, as well as Higgs and Higgsinos are

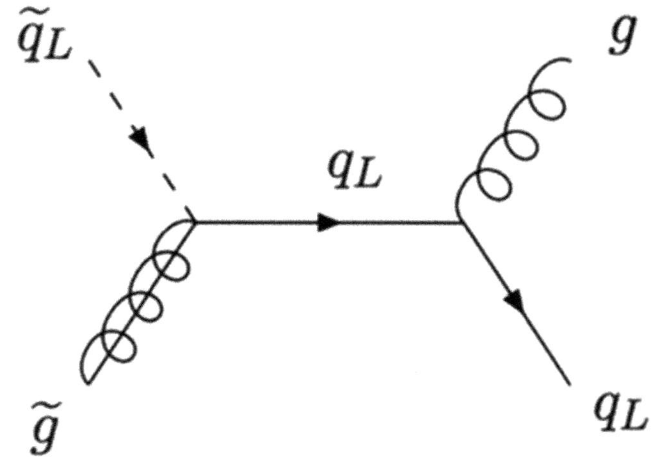

Figure 7.5. Example of a four-body scattering type interaction in the MSSM. When the three-body rates are kinematically forbidden or heavily suppressed the four-body scattering rates dominate. Reprinted with permission from [11],

$$\left\{\Gamma_Y^{\tilde{t}_R,\,\tilde{Q}_L,\,H_1},\,\Gamma_Y^{\tilde{t}_R,\,\tilde{Q}_L,\,H_2},\,\Gamma_Y^{\tilde{t}_R,\,Q_L,\,\tilde{H}},\,\Gamma_Y^{t_R,\,Q_L,\,H_1},\,\Gamma_Y^{t_R,\,Q_L,\,H_2},\,\Gamma_Y^{t_R,\,\tilde{Q}_L,\,\tilde{H}}\right\} = 2y_t^2 N_C$$
$$\left\{|s_\alpha\mu^* + c_\alpha A_t|^2 I_B^{\tilde{t}_R,\,\tilde{Q}_L,\,H_1},\,|c_\alpha\mu^* - s_\alpha A_t|^2 I_B^{\tilde{t}_R,\,\tilde{Q}_L,\,H_2},\,I_F^{\tilde{H},\,\tilde{t}_R,\,Q_L},\,c_\alpha^2 I_F^{t_R,\,Q_L,\,H_1},\right. \tag{7.85}$$
$$\left. s_\alpha^2 I_F^{t_R,\,Q_L,\,H_2},\,I_F^{t_R,\,\tilde{H},\,\tilde{Q}_L}\right\}.$$

7.2.5 Four-body interactions

Four-body scattering processes such as the one depicted in figure 7.5 are typically small compared to the three-body rates calculated in the previous subsection. However, there are regions of parameter space where a three-body rate is kinematically forbidden and the four-body rate dominates. Indeed, if all relevant three-body rates are kinematically suppressed, then ignoring four-body scattering processes would result in particle number densities not relaxing in the symmetric phase. An estimate of the four-body rate for the four-body scattering process involving right-handed tops, Higgs, and a third-generation left-handed quark doublet is [12]

$$\Gamma_{\text{scatt}}^{t,\,H,\,Q} = 0.129\frac{g_3^2}{4\pi}T^2. \tag{7.86}$$

7.3 Local equilibrium approximations

If an interaction is very fast compared to an appropriate time scale (the diffusion time) then one can simplify one's set of transport equations by setting the linear combination of chemical potentials involved in the interaction to zero. Also, if all decay rates from a strongly sourced particular particle species are very slow, you can decouple their transport equations. These two types of steps are the essence of most of the approximations made to make the problem of describing the evolution of particle number densities during the electroweak phase transitions. We will review the most commonly used ones here.

7.4 Gauge and supergauge equilibrium

Gauge interactions are typically assumed to be fast enough to equilibrate gauge multiplets. For example consider an $\mathrm{SU}(2)_L$ doublet denoted by $(\phi_\uparrow, \phi_\downarrow)^T$. If the rate of this interaction with W bosons is sufficiently fast, then the W boson chemical potential is equal to the $\mathrm{SU}(2)_L$ triplet

$$\mu_W = \mu_{\phi_\uparrow} - \mu_{\phi_\downarrow}. \tag{7.87}$$

Moreover, every $\mathrm{SU}(2)_L$ triplet is equal to each other since they are equal to the W boson. Usually the W boson is assumed to have null chemical potential since we are mostly concerned with the dynamics of the symmetric phase near the bubble wall. This is because electroweak sphalerons, which are responsible for producing the baryon asymmetry, are only active in the symmetric phase and the chemical potentials of all $\mathrm{SU}(2)_L$ gauge bosons are zero by symmetry in that phase. This results in all gauge multiplet chemical potentials being equal to zero. Furthermore, gauge triplets can be decoupled from the transport equations as follows. Consider a supergauge interaction, such as the ones depicted in figure 7.6. The source with the left-handed (s)bottom and (s)top is given by

$$\begin{aligned} \partial_\mu j^\mu_{t_L} &\sim g_2^2 |V_{tb}|^2 I_F^{t_L,\, \tilde{b}_L,\, \widetilde{W}^\pm} (\mu_{t_L} - \mu_{\tilde{b}_L} - \mu_{\widetilde{W}^\pm}) \\ \partial_\mu j^\mu_{b_L} &\sim g_2^2 |V_{tb}|^2 I_F^{b_L,\, \tilde{t}_L,\, \widetilde{W}^\pm} (\mu_{b_L} - \mu_{\tilde{t}_L} - \mu_{\widetilde{W}^\pm}). \end{aligned} \tag{7.88}$$

The $\mathrm{SU}(2)_L$ singlet and triplet transport equations are then

$$\begin{aligned} \partial_\mu (j_{t_L} \pm j_{b_L})^\mu = &- \frac{N_C g_2^2}{2} |V_{tb}|^2 \left[\Gamma^{t_L,\, \tilde{b}_L,\, \widetilde{W}} \pm \Gamma^{b_L,\, \tilde{t}_L,\, \widetilde{W}} \right] \\ &\left[(\mu_{t_L} + \mu_{b_L}) - (\mu_{\tilde{t}_L} + \mu_{\tilde{b}_L}) \right] \\ &- \frac{N_C g_2^2}{2} |V_{tb}|^2 \left[\Gamma^{t_L,\, \tilde{b}_L,\, \widetilde{W}} \mp \Gamma^{b_L,\, \tilde{t}_L,\, \widetilde{W}} \right] \\ &\left[(\mu_{t_L} - \mu_{b_L}) - (\mu_{\tilde{t}_L} - \mu_{\tilde{b}_L}) - 2\mu_{\widetilde{W}} \right]. \end{aligned} \tag{7.89}$$

Note that if $\Gamma^{t_L,\, \tilde{b}_L,\, \widetilde{W}} \approx \Gamma^{b_L,\, \tilde{t}_L,\, \widetilde{W}}$ the gauge single transport equation decouples from the dynamics of the gauge triplet. This suppresses the effects of gauge triplets further to the arguments made before and we can, in general, just consider the network of transport equations for gauge singlets. Finally, if supergauge rates such as the ones depicted in figure 7.6 are sufficiently fast, then particles are in equilibrium with their superpartners and we can treat them as one species. Even if they are not, the set of

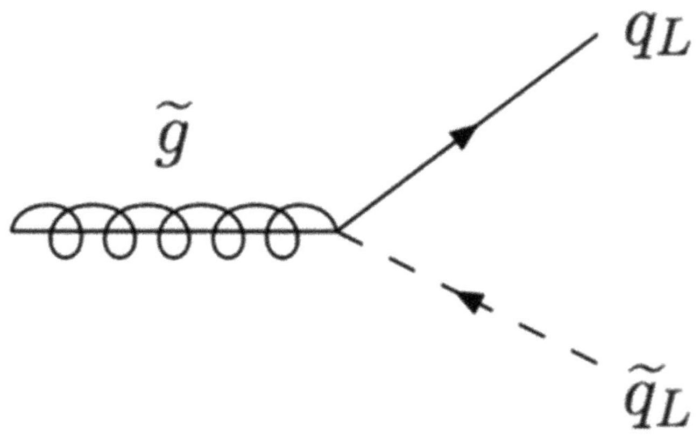

Figure 7.6. Example of a supergauge interaction in the MSSM, or indeed any supersymmetric extension of the Standard Model. Reprinted with permission from [6], copyright (2005) by the American Physical Society.

three-body interactions suppress deviations from supergauge equilibrium. To see this consider the case where all three-body rates are in equilibrium. In such a case if we treat left-handed doublets as a single species (denoted by Q and Q for the (s) quark doublet) we have

$$\begin{aligned} &\mu_Q + \mu_{H_1} - \mu_t = \mu_Q + \mu_{H_2} - \mu_t = \mu_{\tilde{Q}} + \mu_{H_1} - \mu_{\tilde{t}} \\ = \ &\mu_{\tilde{Q}} + \mu_{H_2} - \mu_{\tilde{t}} = \mu_Q + \mu_{\tilde{H}} - \mu_{\tilde{t}} = \mu_{\tilde{Q}} + \mu_{\tilde{H}} - \mu_t \\ = \ &0. \end{aligned} \tag{7.90}$$

Simple manipulation of the above equations then implies supergauge equilibrium even if the supergauge rates are not particularly fast,

$$\mu_Q = \mu_{\tilde{Q}} \quad \mu_t = \mu_{\tilde{t}} \quad \mu_{H_1} = \mu_{H_2} = \mu_{\tilde{H}}. \tag{7.91}$$

So typically, both Higgs doublets and the Higgsino are treated as a single-particle species and gauge singlets and their superpartners are treated as a single-particle species.

7.5 Fast rate approximations

The first two generations of baryons are very weakly sourced as their Yukawa couplings with the Higgs are small. This means there is an approximate baryon number conservation (up to weak sphaleron interactions) for the first two generations. The consequence of this is we can write

$$\mu_{Q_iL} = -2\mu_{d_iR} = -2\mu_{u_iR} \quad \left\{i \in 1,2\right\}. \tag{7.92}$$

By contrast top Yukawa couplings are large, so it is common to approximate that the one-loop Yukawa rate equilibrates the linear combination of chemical potentials

$$\mu_Q = \mu_T - \mu_H. \tag{7.93}$$

Also, the strong sphaleron rate is very large, about a thousand times larger than the weak sphaleron rate. This has inspired people to make the approximation that

$$\mu_5 = \mu_L - \mu_R = 2\mu_{Q_iL} - \mu_{d_iR} - \mu_{u_iR} \quad \left\{i \in (1,2,3)\right\} \approx 2\mu_Q - \mu_T + 9\mu_b = 0, \quad (7.94)$$

where we have used local baryon number conservation for the first two generations. These two approximations can greatly simplify our set of transport equations. Unfortunately, they can give an estimate of the baryon asymmetry that can differ by up to two orders of magnitude from a more precise calculation [10], so we present and later use these approximations for pedagogical purposes only before giving a detailed analytic approach that does not use these approximations.

References

[1] Riotto A 1998 More relaxed supersymmetric electroweak baryogenesis *Phys. Rev.* D **58** 095009

[2] Huet P and Nelson A E 1996 Electroweak baryogenesis in supersymmetric models *Phys. Rev.* D **53** 8

[3] Carena M, Quiros M, Seco M and Wagner C E M 2003 Improved results in supersymmetric electroweak baryogenesis *Nucl. Phys.* B **650** 1

[4] Levinson E J and Boal D H 1985 Self-energy corrections to fermions in the presence of a thermal background *Phys. Rev.* D **31** 3280

[5] Kalashnikov O K 1999 Fermi spectra and their gauge invariance in hot and dense abelian and non-abelian theories *Phys. Scri.* **60** 131

[6] Lee C, Cirigliano V and Ramsey-Musolf M J 2005 Resonant relaxation in electroweak baryogenesis *Phys. Rev.* D **71** 7

[7] Cirigliano V *et al* 2010 Flavored quantum Boltzmann equations *Phys. Rev.* D **81** 10

[8] Weldon H A 1989 Dynamical holes in the quark-gluon plasma *Phys. Rev.* D **40** 2410

[9] Weldon H A 2000 Structure of the quark propagator at high temperatures *Phys. Rev.* D **61** 036003

[10] Cirigliano V, Ramsey-Musolf M J, Tulin S and Lee C 2006 Yukawa and triscalar processes in electroweak baryogenesis *Phys. Rev.* D **73** 115009

[11] Chung D J H, Garbrecht B, Ramsey-Musolf M J and Tulin S 2009 Supergauge interactions and electroweak baryogenesis *JHEP* **12** 067

[12] Joyce M, Prokopec T and Turok N 1996 Nonlocal electroweak baryogenesis. I. Thin wall regime *Phys. Rev.* D **53** 2930

Chapter 8

Plasma and bubble dynamics

Thus far we have neglected thermal masses, widths, diffusion coefficients and the bubble wall velocity. The dynamics of the phase transition and the baryon asymmetry all depend upon these parameters. It is the calculation of these parameters that we focus on in this chapter. Many of these calculations are more conveniently calculated in the finite temperature imaginary time formalism, which is appropriate for systems in equilibrium. Corrections to these parameters due to the departure from equilibrium is assumed to be insignificant. We therefore begin this section with a brief summary of the imaginary time formalism.

8.1 Imaginary time formalism

Fortunately our treatment of the imaginary time formalism can be fairly shallow if our goal is just to give the minimum background necessary to make calculations in this section self-contained. We will therefore merely derive momentum space propagators for both bosons and fermions. Recall that for a system in thermal equilibrium the density matrix is just $Z^{-1}e^{-\beta\mathcal{H}}$. The expectation value of any operator $\mathcal{O}$ is then

$$\langle\mathcal{O}\rangle = Z^{-1}\,\mathrm{Tr}\,[\rho\mathcal{O}]. \tag{8.1}$$

Consider a two point correlator [1]

$$\begin{aligned}\langle\phi(t)\phi(t')\rangle &= Z^{-1}\,\mathrm{Tr}\,[e^{-\beta H}\phi(t)\phi(t')]\\ &= Z^{-1}\,\mathrm{Tr}\,[e^{-\beta H}\phi(t)e^{\beta H}e^{-\beta H}\phi(t')]\\ &= Z^{-1}\,\mathrm{Tr}\,[\phi(t+i\beta)e^{-\beta H}\phi(t')]\\ &= \langle\phi(t')\phi(t+i\beta)\rangle.\end{aligned} \tag{8.2}$$

This identity is known as the Kubo–Martin–Schwinger (KMS) relation [2, 3]. It is a statement that there is an inherent periodicity in the Green's functions at finite temperature that arises from the fact that the density matrix is the time evolution

doi:10.1088/978-0-7503-3571-3ch8

operator for imaginary time β. Note that the above equations implied that the derivation was for bosons but no step assumed this and it applies equally to fermions. Let us Wick rotate to imaginary time. The two-point Green's functions therefore obey the periodicity relation for $\tau > 0$

$$G_\beta(0, \tau) = \pm G_\beta(\beta, \tau). \tag{8.3}$$

A Green's function that is periodic in a space or time variable has a Fourier transform that is a sum over a discrete set of frequencies. We can therefore write for frequencies $\omega_n = (n + n_{\mathrm{F/B}})\pi/\beta$, where $n_{\mathrm{F/B}} = (1, 0)$ for fermions and bosons, respectively [4],

$$G_\beta(\tau) = \frac{1}{\beta}\sum_n \mathrm{e}^{-\mathrm{i}\omega\tau} G_\beta(\omega_n) \tag{8.4}$$

$$G_\beta(\omega_n) = \frac{1}{2}\int_\beta^{-\beta} \mathrm{d}\tau \mathrm{e}^{\mathrm{i}\omega\tau} G_\beta(\tau). \tag{8.5}$$

These frequencies are known as the Matsubara frequencies. The fact that imaginary time becomes periodic at finite temperature is a manifestation of the fact that the temperature that defines the field is not Lorentz invariant. The momentum space Green's functions for bosons and fermions are then

$$\begin{aligned} G_\beta(\vec{k}, \omega_n) &= \frac{1}{\omega_n^2 + \mathrm{k}k^2 + m^2} \\ S_\beta(\vec{k}, \omega_n) &= \frac{\gamma^0\omega_n + \vec{\gamma}\cdot\vec{k} - m}{\omega_n^2 + k^2 + m^2}, \end{aligned} \tag{8.6}$$

respectively.

8.2 Diffusion coefficients

Consider a particle in a system that is out of equilibrium whose number density can be denoted by the function $n(x,t)$. Suppose at a given time $t + \mathrm{d}t$ a particle is at position x, then at a time t the particle must be at the position $x - \mathrm{d}x$. In other words [5]

$$\begin{aligned} n(t + \mathrm{d}t, x) &= \langle n(t, x - \mathrm{d}x)\rangle \\ &= n(t, x) - \frac{\partial n}{\partial x}\langle \delta x\rangle + \frac{1}{2}\frac{\partial^2 n}{\partial x^2}\langle \delta x^2\rangle. \end{aligned} \tag{8.7}$$

Let us assume that our system is isotropic so $\langle \delta x\rangle = 0$. The above equation reduces to

$$\frac{\partial n}{\partial t} = D\frac{\partial^2 n}{\partial x^2} \tag{8.8}$$

for some diffusion coefficient D. The particle number current $J_\mu = (n, -D\vec{\nabla} n)$ implicitly defines the diffusion coefficient via the proportionality of the three-current $\vec{J}$ being proportional to the gradient of the number density[1]

$$\vec{J} = -D\vec{\nabla} n. \tag{8.9}$$

The three-current density can just be defined in the usual way using the distribution function, f, to define the average of the velocity

$$\begin{aligned}\vec{J} = \int \mathrm{d}^3 p f(p)\vec{v} &= \int \mathrm{d}^3 p (f_0 + \delta f)\vec{v} \\ &= \int \mathrm{d}^3 p \delta f \vec{v} \\ &= \int \mathrm{d}^3 p \delta f \frac{\vec{p}}{E}.\end{aligned} \tag{8.10}$$

Here $f_0 = (\exp[\beta(p_0 - \mu)] \pm 1)^{-1}$ are the equilibrium distribution functions for chemical potential μ and fermions (+) and bosons (−). Let us calculate the diffusion coefficients for fermions only. We can relate the number density to the chemical potentials as follows (see chapter 10 for details):

$$n \approx n_0 - \mu \frac{T^2}{12}, \tag{8.11}$$

with n_0 the equilibrium number density. So to find the diffusion coefficient we ultimately need to find a way of solving for δf as well as the chemical potentials. Let us make use of classical Boltzmann equations, in this case the static one which sets $\frac{\partial f}{\partial t} = 0$

$$\frac{\mathrm{d}f}{\mathrm{d}t} = \frac{\vec{p}}{p_0} \cdot \vec{\nabla} f = -C(f). \tag{8.12}$$

We assume that $\vec{\nabla} f \to \dfrac{\partial f}{\partial x}$ and rewrite the left-hand side of the above equation as [7–11]

$$T \frac{p_x}{p_0} \frac{\partial \mu}{\partial x} \frac{\partial f}{\partial(\beta[p_0 - E])}. \tag{8.13}$$

This gives us a potential avenue to relate the chemical potential (or rather its derivative) to other calculable quantities. Let us initially concentrate on calculating the diffusion coefficients of gauge bosons. The collision integral is defined as a sum of four-legged fermionic scattering diagrams with gauge bosons as intermediate states

$$\begin{aligned}C(f) = \int & \frac{\mathrm{d}^3 k \mathrm{d}^3 p' \mathrm{d}^3 k'}{(2\pi)^9 8 E_k E_{p'} E_{k'}} \mid M(s, t, u)|^2 (2\pi)^2 \delta(p + k - p' - k') \\ & \times \left[f_p f_k (1 - f_{p'})(1 f_{k'}) - f_{p'} f_{k'} (1 - f_p)(1 - f_k)\right].\end{aligned} \tag{8.14}$$

[1] For a nice pedagogical introduction to these concepts see [6].

Here $| M |^2$ is the appropriate modulus squared scattering amplitude and (s,t,u) are the usual Mandelstam variables. Let us ignore the fluctuations of the second fluid setting $\delta f_k = \delta f_{k'} = 0$. Let us also assume that little energy is transferred in a collision so $k \sim k'$ and $p \sim p'$. In this case the collision integral simplifies to

$$C(f) = \int \frac{d^3k d^3p' d^3k'}{(2\pi)^9 8E_k E_{p'} E_{k'}} \mid M(s,\, t,\, u)|^2\, (2\pi)^2 \delta(p + k - p' - k') \times (\delta f_p - \delta f_{p'}) f_k^0 (1 - f_k^0). \tag{8.15}$$

To solve the static Boltzmann equation we assume the solution is of the form

$$\delta f_p = g(p,\, x)\frac{p_x}{p_0}. \tag{8.16}$$

Then, using the approximation that energy transfer is low we can write

$$\delta f_p - \delta f_{p'} \approx g(p,\, x)\frac{(p^2 - p{\cdot}p')}{p^2}. \tag{8.17}$$

Our Boltzmann equation then the takes the form

$$T\frac{\partial\mu}{\partial x}\frac{\partial f}{\partial\beta[p_0 - \mu]} = \frac{-1}{p_0^3}\int \frac{d^3k d^3p' d^3k'}{(2\pi)^9 8E_k E_{p'} E_{k'}} \mid M(s,\, t,\, u)|^2\, (2\pi)^2 \delta(p + k - p' - k') \times\, g(p,\, x) p{\cdot}p'\left[\frac{-e^{\beta k_0}}{(e^{\beta k_0} + 1)^2}\right] \equiv -\, g(p,\, x)\Gamma_t, \tag{8.18}$$

where $g(p,x)$ has been taken out of the integral over (p', k, k'). Next substitute our ansatz into our definition of the three current density. Using the relations that $\vec{\nabla} n \rightarrow \frac{\partial n}{\partial x} \propto \frac{\partial\mu}{\partial x}$ we can eliminate the function $g(p,x)$ by combining the definition of the three-current with our Boltzmann equations simplifying the calculation dramatically

$$\begin{aligned} \int d^3p\delta f\frac{p}{E} &= -\, D\vec{\nabla} n \\ \int d^3 g(p,\, x)\frac{p_x^2}{E^2} &= D\frac{T^2}{12}\frac{\partial\mu}{\partial x} D \\ &= \frac{12}{T^2}\int d^3p g(p,\, x)\frac{p_x^2}{E^2}\left(\frac{\partial\mu}{\partial x}\right)^{-1} \\ &= \frac{12}{T^3}\int d^3p\left(\frac{p_x^2}{E^2}\right)\frac{1}{\Gamma_t}\frac{\partial f}{\partial\beta[E - \mu]}. \end{aligned} \tag{8.19}$$

To work out an example consider the case where we have Standard Model gauge interactions. Assuming $p^2 = k^2 = \cdots = 0$, the relevant modulus squared scattering amplitudes are

$$\begin{aligned} |M_W|^2 &= \frac{144 g_2^4 (p\cdot k)^2}{(2p\cdot p' + M_W^2)^2} \\ |M_B|^2 &= \frac{320 Y^2 g_1^4 (p\cdot k)^2}{(2p\cdot p' + M_B^2)^2} \\ |M_G|^2 &= \frac{128 g_3^4 (p\cdot (k + k'))^2}{(2p\cdot p' + M_B^2)^2}, \end{aligned} \tag{8.20}$$

with thermal masses

$$M_B^2 = \frac{g_2^2 \ \tan^2 \ \theta_W T^2}{3} M_W^2 = \frac{5 g_2^2 T^2}{3} M_G^2 = 2 g_3^2 T^2. \tag{8.21}$$

Let us calculate Γ_t in the center-of-mass frame and calculate only to first-order in M/T. The gluon case gives, for example,

$$\Gamma_t \approx \frac{1}{6\pi} \frac{T^3}{p^2} \log\left(\frac{4pTx_2}{M^2}\right). \tag{8.22}$$

We then have a pretty nasty integral to perform that includes a logarithm in the integrand. We can approximate this integral by taking the logarithm out of the integrand and evaluating it at the point p which maximizes $p^6 \frac{\partial f}{\partial \beta[E - \mu]}$. This gives

$$\begin{aligned} D_B^{-1} &\approx \frac{100}{7\pi} \alpha_W^2 \tan^4 \theta_W Y^2 T \log\left(\frac{32T^2}{M_B^2}\right) \\ D_W^{-1} &\approx \frac{45}{7\pi} \alpha_W^2 T \log\left(\frac{32T^2}{M_W^2}\right) \\ D_G^{-1} &\approx \frac{80}{7\pi} \alpha_S^2 T \log\left(\frac{32T^2}{M_G^2}\right). \end{aligned} \tag{8.23}$$

The diffusion coefficients of fermions in the Standard Model is just then the sum of the contributions from respective gauge bosons

$$\begin{aligned} D_{l_R}^{-1} &= D_B^{-1} \approx \frac{T}{380} \\ D_{l_L}^{-1} &= D_W^{-1} + \frac{1}{4} D_{l_R}^{-1} \approx \frac{T}{100} \\ D_{q_{L,R}}^{-1} &= Y^2 D_{l_R}^{-1} + \epsilon_{L,R} D_W^{-1} + D_G^{-1} \approx \frac{T}{6}, \end{aligned} \tag{8.24}$$

where $\epsilon_{L,R} = (1, 0)$.

8.3 Thermal widths

The poles of a propagator are shifted in zero temperature field theory by an amount equal to the expectation value of imaginary part of the self-energy

$$\Gamma = \langle \mathrm{Im}(\Sigma(P)) \rangle. \tag{8.25}$$

At finite temperature we do the same using finite temperature propagators. The self-energy is defined along the real axis, $\Sigma(P)$. Let $f(P)^{-1} = \Sigma(P)$, the imaginary part of the self-energy is given by [12]

$$\mathrm{Im}\,[\Sigma(P)] = \frac{1}{2i}\left(\frac{1}{f(p + \mathrm{i}\epsilon)} - \frac{1}{f(p - \mathrm{i}\epsilon)}\right). \tag{8.26}$$

There are two types of contribution to this, residues of poles and discontinuities across the real axis [13–15]. In practice the thermal width is dominated by gauge interactions. For the MSSM the thermal mass has been found in various limits. In the limit where fermions have masses that are heavy compared to the temperature the contribution is [16]

$$\Gamma = \frac{g^2 T C_f}{8\pi}, \tag{8.27}$$

where C_f is the Casimir invariant of the fermion representation. In the massless fermion limit at rest, the contribution is the same as the above with the coefficient $2a(N, N_F)$ which depends on the gauge group and the number of fermions. Finally, the $p > 0$ massless fermion limit gives a small velocity-dependent contribution that in general can be ignored. The thermal width of fermions interacting with scalars is generally quite small. For the Higgsino–stop interaction the contribution is $\sim 0.01 T y^2$ in the massless limit and $\lesssim 0.0025 T y^2$ for $M = 0(T/2)$ for scalar mass M and Yukawa coupling y. The thermal width of fermions with gauge symmetry is therefore $\approx \frac{g^2 T C_f}{4\pi}$, where the variation in $a(N, N_F)$ and the contribution from moving massless fermions is ignored. A table of values of all thermal widths relevant to baryon asymmetry calculation in the MSSM is given in table 8.1.

Table 8.1. Thermal widths relevant to the MSSM. Values taken from [16].

Relevant thermal widths in the MSSM	
Particle	Thermal width
$\Gamma_{\tilde{H}}$	0.025 T
$\Gamma_{\tilde{B}}$	0.003 T
$\Gamma_{\tilde{W}}$	0.065 T
Γ_t	0.5 T
$\Gamma_{\tilde{t}}$	0.5 T
Γ_τ	0.002 T
$\Gamma_{\tilde{t}}$	0.002 T

8.4 Thermal masses

We now turn our attention to calculating the one loop thermal correction to the mass, known as the Debye mass [1, 17]. Let us start off by calculating the easiest example. Consider the model

$$\mathcal{L} = \frac{1}{2}\partial_\mu\phi\partial^\mu\phi - \frac{m^2}{2}\phi^2 - \frac{\lambda}{4!}\phi^4. \tag{8.28}$$

In the imaginary time formalism the integral over k_0 is replaced with a sum over Matsubara frequencies. The diagram in figure 8.1 is

$$\Delta m^2 = \frac{\lambda}{2\beta}\sum_n\int\frac{\mathrm{d}^3k}{(2\pi)^3}\frac{1}{\left(\frac{2n\pi}{\beta}\right)^2 + k^2 + m^2}. \tag{8.29}$$

Note that the sum $(n^2 + y^2)^{-1} = (\pi/y)\coth \pi y$, which allows us to write the above as

$$\Delta m^2 = \frac{\lambda}{4}\int\frac{\mathrm{d}^3k}{(2\pi)^3}\frac{1}{\sqrt{k^2 + m^2}}\quad\coth\frac{\beta}{2}\sqrt{k^2 + m^2}. \tag{8.30}$$

We can separate this into $0T$ and thermal corrections using the definition of the hyperbolic trigonometric function $\coth \beta y = 1 + 2n_B(2y)$ and write

$$\Delta m^2 = \frac{\lambda}{4}\int\frac{\mathrm{d}^3k}{(2\pi)^3}\frac{1}{\sqrt{k^2 + m^2}} + \frac{\lambda}{2}\int\frac{\mathrm{d}^3k}{(2\pi)^3}\frac{1}{\sqrt{k^2 + m^2}}\frac{1}{e^{\beta\sqrt{k^2+m^2}} - 1}. \tag{8.31}$$

Let us denote the second term by Δm_T^2. In the limit where $m \to 0$ (or if the mass is small compared to the temperature) then we can make the approximation

$$\begin{aligned}\Delta m_T^2 = \frac{\lambda}{2}\int\frac{d^3k}{(2\pi)^3}\frac{1}{\sqrt{k^2+m^2}}\frac{1}{e^{\beta\sqrt{k^2+m^2}} - 1} &\sim \frac{4\pi\lambda}{2}\int_0^\infty\frac{p^2dp}{(2\pi)^3}\frac{1}{p}\frac{1}{e^{\beta p} - 1},\\ &= \frac{\lambda}{4\pi^2\beta^2}\int_0^\infty\frac{pdp}{e^p - 1},\\ &= \frac{\lambda T^2}{24}.\end{aligned} \tag{8.32}$$

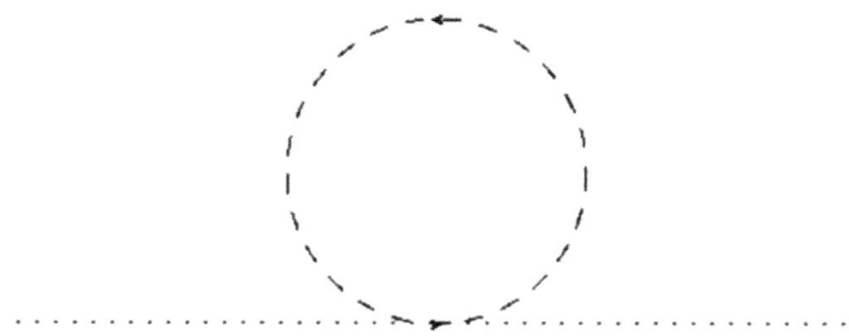

Figure 8.1. Relevant diagram for calculating the thermal mass correction for a scalar field theory with ϕ^4 interactions.

Table 8.2. Relevant thermal masses for electroweak baryogenesis (EWBG) in the MSSM separated into Standard Model contributions (A), interactions with Higgsinos, binos, winos, and RH stops (B) and RH sbottoms and staus (C). Table reproduced from [18].

Thermal masses in the MSSM

Particle	$\delta m_{\rm SM}/T^2$ (A)	$\delta m_{\rm SUSY}/T^2$ (B)	$\delta m_{\rm SUSY}/T^2$ (C)
qL	$\frac{1}{6}g_3^2 + \frac{3}{32}g_2^2 + \frac{1}{288}g_1^2 + \frac{1}{16}y_t^2 + \frac{1}{16}y_b^2$	$+\frac{1}{16}y_t^2$	$+\frac{1}{16}y_b^2$
tR	$\frac{1}{6}g_3^2 + \frac{1}{18}g_1^2 + \frac{1}{8}y_t^2$	$+\frac{1}{18}g_1^2$	
bR	$\frac{1}{6}g_3^2 + \frac{1}{72}g_1^2 + \frac{1}{8}y_b^2$		$+\frac{1}{72}g_1^2$
lL	$\frac{3}{32}g_2^2 + \frac{1}{32}g_1^2 + \frac{1}{16}y_\tau^2$		$\frac{1}{16}y_\tau^2$
τ_R	$\frac{1}{8}g_1^2 + \frac{1}{8}y_\tau^2$		$+\frac{1}{8}g_1^2$
$\tilde{H}_u$		$\frac{3}{16}g_2^2 + \frac{1}{16}g_1^2 + \frac{3}{16}y_t^2$	
$\tilde{H}_d$		$\frac{3}{16}g_2^2 + \frac{1}{16}g_1^2$	$+\frac{3}{16}y_b^2 + \frac{1}{16}y_\tau^2$
$\widetilde{W}$		$\frac{3}{8}g_2^2$	
$\widetilde{B}$		$\frac{5}{12}g_2^2$	$+\frac{1}{6}g_2^2$
$\tilde{t}_R$		$\frac{4}{9}g_3^2 + \frac{1}{3}g_1^2 + \frac{1}{3}y_t^2$	$-\frac{1}{9}g_1^2$
$\tilde{b}_R$			$\frac{4}{9}g_3^2 + \frac{1}{18}g_1^2 + \frac{1}{3}y_b^2$
$\tilde{\tau}_R$			$\frac{1}{2}g_1^2 + \frac{1}{3}y_\tau^2$
Hu	$\frac{3}{16}g_2^2 + \frac{1}{16}g_1^2 + \frac{1}{4}y_t^2$	$\frac{3}{16}g_2^2 - \frac{1}{48}g_1^2 + \frac{1}{4}y_t^2$	$\frac{1}{12}g_1^2$
Hd	$\frac{3}{16}g_2^2 + \frac{1}{16}g_1^2 + \frac{1}{4}y_t^2 + \frac{1}{12}y_\tau^2$	$\frac{3}{16}g_2^2 + \frac{7}{48}g_1^2$	$-\frac{1}{12}g_1^2 + \frac{1}{4}y_b^2 + \frac{1}{12}y_\tau^2$

where in the last line we have used the identity $\int p/(e^p - 1) = \pi^2/6$. In table 8.2 the thermal masses for the MSSM are given, separated into their Standard Model and beyond the Standard Model (BSM) contributions.

8.5 Bubble wall velocity

Calculating the bubble wall velocity is a difficult task and there are essentially two approaches that tend to be used. One approach requires the use of multiple approximations including, at one stage, treating the particles in the plasma classically. This approach is not always viable. For example, if multiple fields have a space–time varying vacuum during a phase transition the approximations we will introduce in this section are not necessarily valid and one needs to follow a purely numerical approach [19–21]. The conventional semi-analytic approach begins by deriving the classical equations of motion [22, 23]. For a single scalar field this is

$$\Box\phi - \frac{\partial V}{\partial\phi} - \sum \frac{dm_A^2}{d\phi}A^2 + \frac{dm_\Psi}{d\phi}\overline{\Psi}_R\Psi_L, \tag{8.33}$$

where we have suppressed gauge indices. We set the scalar field to a classical value with symmetric perturbations. That is, $\phi \to \phi_{\rm cl} + \delta\phi$ with $\langle\delta\phi\rangle = 0$. We then take the trace of the finite temperature density matrix to obtain

$$\Box\phi_{\rm cl} - \frac{\partial V}{\partial\phi}_{\rm cl} - \frac{{\rm d}m_\phi^2}{{\rm d}\phi}\langle\delta\phi^2\rangle - \sum\frac{{\rm d}m_A^2}{{\rm d}\phi}\langle A^2\rangle + \frac{{\rm d}m_\Psi}{{\rm d}\phi}\langle\overline{\Psi}_R\Psi_L\rangle. \tag{8.34}$$

If one assumes Wentzel–Kramers–Brillouin (WKB) waves then one can write the second moment of each field as

$$\begin{aligned}\langle A^2\rangle &= \langle A^2\rangle_{\rm vac} + \sum\int\frac{{\rm d}^3k}{(2\pi)^3E}\, f(k, x)\\ \langle\overline{\Psi}_R\Psi_L\rangle &= \frac{1}{2}\langle\overline{\Psi}_R\Psi_L\rangle_{\rm vac} + \sum\int\frac{{\rm d}^3k}{(2\pi)^3}\frac{m}{2E}\, f(k, x),\end{aligned} \tag{8.35}$$

where we have omitted the group and Dirac indices. Using the trivial identity that $m\frac{{\rm d}m}{{\rm d}\phi} = \frac{1}{2}\frac{{\rm d}m^2}{{\rm d}\phi}$ and omitting the subscript 'cl', we can write the equations of motion in the form

$$\Box\phi - \frac{\partial V}{\partial\phi} + \sum\frac{{\rm d}m^2}{{\rm d}\phi}\int\frac{{\rm d}^3k}{(2\pi)^3 2E} f(k, x) = 0. \tag{8.36}$$

If we multiply both sides by $\partial^\mu\phi$ we obtain a nice physical interpretation where the divergence of the stress energy tensor for the scalar field is equal to the force a particle feels in the presence of a wall,

$$\partial_\nu T^{\nu\mu}(\phi) = \sum\int\frac{{\rm d}^3k}{(2\pi)^3} f(k)\partial^\mu m^2(\phi). \tag{8.37}$$

The particle distribution functions can be assumed to be a correction to their equilibrium counterparts, $f = f_0 + \delta f$, with $f_0 = (\exp[\beta\gamma(E - vp_z)] \pm 1)^{-1}$ defined in the usual way. The equilibrium terms, f_0, are absorbed into the effective potential and become the one-loop temperature corrections $V(\phi) \to V(\phi,T)$. The particle distribution functions are approximated to obey classical Boltzmann equations

$$\begin{aligned}\left(\frac{{\rm d}f}{{\rm d}t}\right)_{\rm collisions} &= \frac{\partial f}{\partial t} + \{f, H\}\\ - \quad C(f) &= \frac{\partial f}{\partial t} + \frac{\partial x}{\partial t}\frac{\partial f}{\partial x} + \frac{\partial p}{\partial t}\frac{\partial f}{\partial p}.\end{aligned} \tag{8.38}$$

Let us calculate the right-hand side explicitly

$$\begin{aligned}- C(\ f) &= \frac{\partial f}{\partial t} - \left[\frac{\partial x}{\partial t}\frac{\beta\gamma}{E}\frac{\partial m^2}{\partial x} - \frac{\partial p}{\partial t}\beta\gamma\left(\frac{p_x}{E} - v\right)\right]\frac{{\rm e}^{\beta\gamma(E-vp_x)}}{({\rm e}^{\beta\gamma(E-vp_x)} \pm 1)^2}\\ &\equiv \frac{\partial f}{\partial t} - \left[\frac{\partial x}{\partial t}\frac{\beta\gamma}{E}\frac{\partial m^2}{\partial x} - \frac{\partial p}{\partial t}\beta\gamma\left(\frac{p_x}{E} - v\right)\right]h_{F,\,B}(\beta,\, E,\, p_x).\end{aligned} \tag{8.39}$$

If we take our frame of reference to be the bubble wall rest frame then $\delta f = \delta f(x + v_w t)$ with v_w the wall velocity. This implies that $\frac{\partial \delta f}{\partial p} = 0$, $\frac{\partial x}{\partial t} = \frac{p_x}{E}$ and $\frac{\partial p_x}{\partial t} = \frac{-1}{2E}\frac{\partial m^2}{\partial x} v_x$. Furthermore, the Boltzmann equations are now reduced to

$$\frac{p_x}{E}\frac{\partial \delta f}{\partial x} - \frac{\partial m^2}{\partial x}\frac{v\beta\gamma}{2E} h_{F,B}(\beta, E, p_z) = -C(f). \tag{8.40}$$

Finally, let us define the collision integral. For four-body collisions we sum over all four-legged scattering diagrams evaluated at finite temperature, which means

$$\begin{aligned} C(f) = \sum \frac{1}{2E_p} \int \frac{d^3k d^3p' d^3k'}{(2\pi)^9 8 E_k E_{p'} E_{k'}} |M(s, t)|^2 (2\pi)^2 \delta(p + k - p' - k') \\ \times \left[f_1 f_2 (1 \pm f_3)(1 \pm f_4) - f_3 f_4 (1 \pm f_1)(1 \pm f_2) \right], \end{aligned} \tag{8.41}$$

where the sum of course is over diagrams. We can solve these Boltzmann equations exactly for the case when the collision integral is zero, which applies in the case where the mean free path is greater than the bubble wall width,

$$f^{-1} = \exp\left[\frac{E + v_w}{(1 - v_w^2)T} + s_i \frac{\gamma v_w}{T} \left(\frac{(p_x + v_w E)^2}{1 - v_w^2} + m_i^2 \right)^{\frac{1}{2}} \right] \pm 1, \tag{8.42}$$

with

$$(m_i, s_i) = \begin{cases} (m^2, -1) & p_x > -\sqrt{m_0^2 - m^2} \\ (m^2 - m_0^2, +1) & p_x < -\sqrt{m_0^2 - m^2} \end{cases} \tag{8.43}$$

and the $\pm$ sign in the solution of f^{-1} is for fermions and bosons, respectively. Going beyond this approximation we can consider two regions, the slow and fast wall regimes. Let the collision integral be $C \approx \frac{\delta f}{L_w}$, where L_w is the wall width. In the slow wall regime we assume that $\delta f' \approx \frac{\delta f}{L} \ll \frac{\delta f}{\tau}$ for a diffusion time τ. In this case we can write

$$\delta f = \tau \frac{\partial m^2}{\partial x}\frac{v\beta\gamma}{2E} h_{F,B}(\beta, E, p_x), \tag{8.44}$$

in which case the equations of motion are

$$\frac{\partial^2 \phi}{\partial x^2} + \frac{\partial V(\phi, T)}{\partial \phi} + \sum \frac{dm^2}{d\phi} \int \frac{d^3p}{(2\pi)^3}\frac{\tau\beta v_w \gamma}{(2E)^2}\frac{\partial m^2}{\partial x} h_{F,B}(\beta, E, p_x) = 0. \tag{8.45}$$

The other region we can assume the fast wall approximation where $\delta f' \approx \frac{\delta f}{L} \gg \frac{\delta f}{\tau}$ we have [24]

$$\frac{p_x}{E}\delta f' = \frac{\partial m^2}{\partial x}\frac{v_w \beta\gamma}{2E}h_{F,\,B}(\beta,\,E,\,p_x) \tag{8.46}$$

in which case the equations of motion are

$$\frac{\partial^2\phi}{\partial x^2} + \frac{\partial V(\phi,\,T)}{\partial\phi} + \sum\frac{\mathrm{d}m^2}{\mathrm{d}\phi}\int\frac{\mathrm{d}^3p}{(2\pi)^3}\frac{m^2 v_w\beta\gamma}{4Ep_x}h_{F,\,B}(\beta,\,E,\,p_x) = 0. \tag{8.47}$$

In both cases the equations have to be solved numerically.

References

[1] Kapusta J I and Landshoff P V 1989 Finite temperature field theory *J. Phys. G: Nucl. Part. Phys.* **15** 267

[2] Kubo R 1957 Statistical-mechanical theory of irreversible processes. I. General theory and simple applications to magnetic and conduction problems *J. Phys. Soc. Japan* **12** 570

[3] Martin P and Schwinger J 1959 Theory of many-particle systems. I *Phys. Rev.* **115** 1342

[4] Matsubara T 1955 A new approach to quantum-statistical mechanics *Prog. Theor. Phys.* **14** 351

[5] Fick A 1855 Ueber diffusion *Ann. Phys.* **94** 59

[6] Tong D 2012 *Kinetic Theory: University of Cambridge Graduate Course* (Cambridge: University of Cambridge)

[7] Joyce M, Prokopec T and Turok N 1996 Nonlocal electroweak baryogenesis. I. Thin wall regime *Phys. Rev.* D **53** 2930

[8] Joyce M, Prokopec T and Turok N 1995 Electroweak baryogenesis from a classical force *Phys. Rev. Lett.* **75** 9

[9] Arnold P, Moore G D and Yaffe L G 2003 Transport coefficients in high temperature gauge theories, 2. Beyond leading log *J. High Energy Phys. JHEP05* **2003** 051

[10] Arnold P, Moore G D and Yaffe L G 2001 Photon emission from quark-gluon plasma: complete leading order results *J. High Energy Phys. JHEP12* **2001** 009

[11] Joyce M, Prokopec T and Turok N 1996 Nonlocal electroweak baryogenesis. II. The classical regime *Phys. Rev.* D **53** 6

[12] Kobes R, Kunstatter G and Mak K 1992 Fermion damping in hot gauge theories *Phys. Rev.* D **45** 12

[13] Kobes R L and Semenoff G W 1985 Discontinuities of Green functions in field theory at finite temperature and density *Nucl. Phys.* B **260** 3

[14] Pisarski R D 1989 Scattering amplitudes in hot gauge theories *Phys. Rev. Lett.* **63** 11

[15] Pisarski R D 1993 Damping rates for moving particles in hot, QCD *Phys. Rev.* D **47** 12

[16] Elmfors P *et al* 1999 Damping rates in the MSSM and electroweak baryogenesis *Phys. Lett.* B **452** 3

[17] Kapusta J I and Landshoff P V 1989 Finite-temperature field theory *J. Phys. G: Nucl. Part. Phys.* **15** 3

[18] Chung D J H *et al* 2010 Lepton-mediated electroweak baryogenesis *Phys. Rev.* D **81** 6

[19] Moore G D and Prokopec T 1995 How fast can the wall move? A Study of the electroweak phase transition dynamics *Phys. Rev.* D **52** 7182

[20] John P and Schmidt M G 2001 Do stops slow down electroweak bubble walls? *Nucl. Phys.* B **598** 291

[21] Kozaczuk J 2015 Bubble expansion and the viability of singlet-driven electroweak baryogenesis *J. High Energy Phys.* **10** 1–46
[22] Moore G D and Prokopec T 1995 Bubble wall velocity in a first order electroweak phase transition *Phys. Rev. Lett.* **75** 5
[23] Moore G D 2000 Electroweak bubble wall friction: analytic results *J. High Energy Phys. JHEP03* **2000** 006
[24] Huber S J and Sopena M 2013 An efficient approach to electroweak bubble velocities arXiv: hep-ph-1302.1044

IOP Publishing

Electroweak Baryogenesis (Second Edition)
An introduction
Graham White

Chapter 9

Perturbation theory and resummation at finite temperature

There is a serious problem that we have so far all but neglected in our treatment of the phase transitions—perturbation theory breaks down badly at finite temperature. This will mean our prediction for the critical and nucleation temperature will be badly inaccurate. Worse, the gravitational wave peak amplitude scales as

$$\Omega_{GW} \sim \alpha^2/R_*^2. \tag{9.1}$$

Here α is, to a good approximation, a ratio of the difference in pressure—which is expected to scale inversely with the fourth power of the temperature—and the energy density for radiation domination. The mean bubble separation probably has an exponential scaling for most models. However, near enough to a given temperature, the scaling is linear. Therefore, if there is some uncertainty in the temperature at which the gravitational waves are produced, the peak amplitude scales with the 18th power of that temperature. So, the large uncertainty from the breakdown of perturbation theory at high temperature renders the framework barely predictive at all when it comes to gravitational waves.

The reason for this breakdown comes from the fact that our expansion parameter at finite temperature is no longer the coupling constant divided by a phase space factor, but contains the mode occupation [1]

$$\frac{g^2}{\pi} \to \frac{g^2}{\pi} n_b \sim \frac{g^2}{\pi}\frac{T}{m}, \tag{9.2}$$

where in the last line we used the high temperature expansion for $n_b = (\exp[\beta E] - 1)^{-1}$. Consider that masses are field dependent, and we are concerned about the value of the effective potential for all values of the field between the true and false vacuum. If the mass becomes very small compared to the temperature, or even vanishes, then the expansion parameter can quickly become

doi:10.1088/978-0-7503-3571-3ch9

large. The problem is less awful for fermions, which lack an infrared divergence. However, even for fermions, the mode occupation reaches its maximum value when the mass vanishes, indicating that they can also slow down the convergence of perturbation theory.

The traditional approach to rescuing perturbation theory is to resume loop corrections to the mass such that

$$\frac{g^2}{\pi}\frac{T}{m} \to \frac{g^2}{\pi}\frac{T}{\sqrt{m^2 + \delta m_T^2}}, \tag{9.3}$$

which superficially seems to solve the issue. In the case of the Higgs and Goldstone modes, however, the square of the mass before resummation is negative. This means that for small field values, adding the thermal mass correction, δm_T^2, actually makes the breakdown of perturbation theory worse. However, an exact cancellation only happens for a second-order transition. For a first-order transition, the mass remains positive by virtue of the fact that we have a global minimum at the origin. Thus we can conjecture that resummation is a program that could help save perturbation theory at high temperature in the phenomologically interesting cases where the transition is sufficiently strongly first order.

In the $\overline{MS}$ scheme that we used to derive the effective potential in chapter 4, the potential as well as all physical observables we derive from it are dependent upon an RG scale. This scale dependence is an artifact of the finite truncation of perturbation theory. The parameters in the tree-level potential have an implicit scale dependence which partially cancels against the explicit scale dependence in the one-loop Coleman–Weinberg potential. A full cancellation occurs up to the order in our loop expansion when one includes the wave function renormalization term, any residual dependence is suppressed by higher powers of the coupling and phase space suppression. A test of the success of perturbation theory then at finite temperature would be to check if, after resummation, the scale dependence cancels at the same order the theory is defined at.

Let us consider the case of a pure scalar theory with a discrete symmetry, the simplest theory that can undergo a thermal phase transition. This example is quite clean, as it has the advantage of the trivial wave function renormalization meaning the implicit scale dependence should cancel the explicit dependence that appears in the loop functions. The effective potential at tree-level has the form

$$V_{\text{tree}} = \frac{1}{2}m^2\varphi^2 + \frac{g^2}{4!}\varphi^4. \tag{9.4}$$

The Coleman–Weinberg potential contains a single term

$$V_{\text{CW}} = \frac{M^4}{64\pi^2}\left(\log\left[\frac{M^2}{\mu^2}\right] - \frac{3}{2}\right), \tag{9.5}$$

where μ is the RG scale in the $\overline{MR}$ scheme and M is the field dependent mass satisfying, $M^2 = m^2 + \frac{g^2}{2}\varphi^2$. We can glean the implicit scale dependence of the tree-level potential by writing the renormalization group equations

$$\mu\frac{dm^2}{d\mu} = \frac{g^2m^2}{(4\pi)^2}, \tag{9.6}$$

$$\mu\frac{dg^2}{d\mu} = \frac{3g^4}{(4\pi)^2}. \tag{9.7}$$

The scale dependence of the tree and loop contributions to the potential, up to $O(g^4)$, are as follows

$$\mu\frac{dV_{\text{tree}}}{d\mu} = \frac{1}{2}\frac{g^2m^2}{(4\pi)^2}\varphi^2 + \frac{3g^4}{24(4\pi)^2}\varphi^4, \tag{9.8}$$

$$\mu\frac{dV_{\text{CW}}}{d\mu} = -\frac{M^4}{32\pi^2} = -\frac{1}{32\pi^2}\left(m^4 + m^2g^2\varphi^2 + \frac{1}{4}g^4\varphi^4\right). \tag{9.9}$$

This cancels up to an irrelevant field independent term which can be cancelled by setting the potential to zero at the origin of field space. Next let us consider finite temperature corrections. We will work in the high temperature expansion to make the problem analytically transparent,

$$V_T = \frac{g^2}{48}T^2\varphi^2 - \frac{1}{12\pi}TM^{3/2} - \frac{M^4}{64\pi^2}L_b(\mu). \tag{9.10}$$

Here $L_b(\mu) = 2\log\left(\frac{e^{\gamma_E}\mu}{4\pi T}\right)$ and $\gamma_E = 0.577$ is the Euler–Maclaurin constant. Let us focus on the quadratic term first (the other term we will call a barrier term for lack of a better word, since there isn't much of a barrier in scalar ϕ^4 theory. However, this effective cubic term, is the term that generally produces a barrier between the minima in a more complicated theory). The scale dependence is

$$\mu\frac{dV_T}{d\mu} = \frac{3}{768\pi^2}g^2T^2\varphi^2. \tag{9.11}$$

There is nothing to cancel this term which is the same order as terms arising from the tree-level potential, yet this is a one-loop term. This indicates the breakdown in perturbation theory prior mentioned. Note that the breakdown in perturbation theory is, as expected, worse in the infrared when the field value is small as the ratio of this term to other terms arising from the tree-level potential diverges. Secondly, we can observe that some one-loop corrections are the same size as tree-level corrections, some two-loop corrections are the same size as one-loop corrections and so on. So we need to reorganize our calculation to handle the mismatch as well as removing the divergences. To do so, we need a size measure to discern how large a diagram is—the power of the coupling presents a good candidate for this. So we need to reorganize our perturbation theory so that it is correctly defined to at least $O(g^4)$.

Let us then see the application of the simplest resummation scheme suggested before. In this case the mass in the propagator is shifted and all loop functions will have a shifted mass

$$V_{1-\text{loop}} = V_{\text{tree}} + V_{\text{CW}}(M^2 + \delta m^2) + V_T(M^2 + \delta m^2).$$

This method is known as the Parwani resummation scheme [2]. The calculation of the thermal mass we calculated in chapter 8, and it involves the inclusion of the below diagram. Here we will just quote the result.

The scale dependence of the Coleman–Weinberg term now partially cancels the thermal quadratic term

$$\mu\frac{\mathrm{d}V_{\text{CW}}}{\mathrm{d}\mu} = -\frac{(m^2 + \frac{g^2T^2}{24} + \frac{g^2\varphi^2}{2})^2}{32\pi^2}. \tag{9.12}$$

This cancellation demonstrates that resummation has improved the convergence of our perturbation theory. To fully cancel the scale dependence at this order, one also needs to include the lollypop diagram.

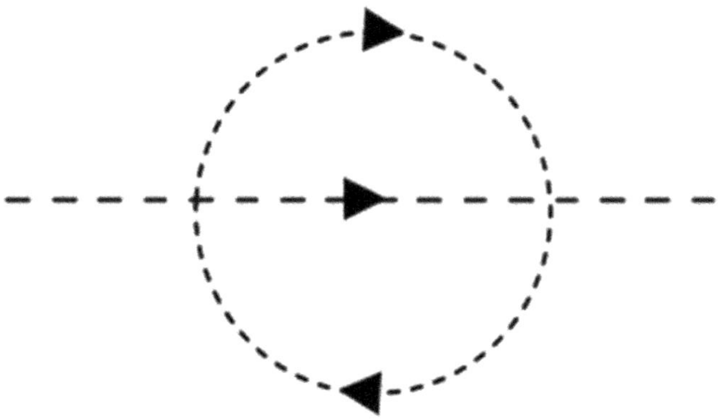

This diagram can be approximated as

$$\sim 2\frac{g^4\phi^2T^2}{768\pi^2}\left(\log\left[\frac{M^2}{T\mu}\right] + 1.65\right), \tag{9.13}$$

Which is exactly what is needed to cancel the scale dependence at this order. So far this looks reasonably tractable—just perform the above resummation and then add the missing diagrams by hand in order to make the calculation accurate to a certain power of the couplings. Of course, the scale dependence of the barrier term is completely untreated in the Parwani scheme. An estimate of how important the neglect of such a barrier term can be discerned by calculating,

$$\mu\frac{\mathrm{d}V_{\text{barrier}}}{\mathrm{d}\mu} = \frac{MT}{192\pi^3}\left(m^2g^2 + \frac{3}{2}g^4x^2\right). \tag{9.14}$$

This is suppressed by extra powers of pi, but in the limit where $M \sim gT$, is of $O(g^3/\pi^3)$, which is not necessarily that suppressed compared to the terms of $O(g^4/\pi^2)$. In principle one can add higher loop diagrams, one at a time to cancel this term with care to avoid double counting. Unfortunately, while this looks plausible in φ^4 theory, things quickly become intractably messy without a framework to organize the problem. To see this, let us consider the Standard Model with just top quarks having a non-zero Yukawa coupling. The renormalization group equations are slightly more complicated

$$\mu\frac{\mathrm{d}m^2}{\mathrm{d}\mu} = \left(-\frac{3}{2}g_1^2 - \frac{9}{2}g_2^2 + 12\lambda + 6y_t^2\right)\frac{m^2}{16\pi^2}, \tag{9.15}$$

$$\mu\frac{\mathrm{d}g_1^2}{\mathrm{d}\mu} = \frac{41}{48\pi^2}\frac{g_1^4}{}, \tag{9.16}$$

$$\mu\frac{\mathrm{d}g_2^2}{\mathrm{d}\mu} = \frac{41}{48\pi^2}\frac{g_2^4}{}, \tag{9.17}$$

$$\mu\frac{\mathrm{d}\lambda}{\mathrm{d}\mu} = \left(\lambda[-3g_1^2 - 9g_2^2 + 12y_t^2] + \frac{3}{8}(g_1^4 + 3g_2^4 + 2g_1^2g_2^2) + 24\lambda^2 - 6y_t^2\right)\frac{1}{16\pi^2}, \tag{9.18}$$

$$\mu\frac{\mathrm{d}y_t^2}{\mathrm{d}\mu} = \left(3y_t^4 + 2y_t^2 + \left[-\frac{17}{20}g_1^2 - \frac{9}{4}g_2^2 - 8g_3^2 + 3y_t^2\right]\right)\frac{1}{16\pi^2}. \tag{9.19}$$

In this case, taking the tree-level effective potential to be $V_{\text{tree}}(h) = \frac{m^2}{2}h^2 + \frac{\lambda}{4}h^4$, in this case the scale dependence of the Coleman–Weinberg potential has the form

$$\mu\frac{\mathrm{d}V_{\text{CW}}}{\mathrm{d}\mu} = -\frac{1}{32\pi^2}(m_H^4 + 3m_G^4 + 6m_W^4 + 3m_z^4 - 12m_t^4), \tag{9.20}$$

where the field dependent masses of the Standard Model in the Landau gauge have the form

$$m_h^2 = m^2 + 3\lambda h^2,\quad m_G^2 = m^2 + \lambda h^2,\quad m_W^2 = \frac{g_2^2}{4},\quad m_Z^2 = \frac{g_1^2 + g_2^2}{4},\quad m_t^2 = \frac{y_t^2}{2}. \tag{9.21}$$

A simple calculation shows that at zero temperature, the scale dependence of the Coleman–Weinberg piece does not cancel the scale dependence of the tree-level potential. Nor should it, as the wave function renormalization in the Standard model is non-trivial. In fact, the full Callan–Symanzik equation to first order in the loop expansion has the form

$$\mu\frac{\mathrm{d}V}{\mathrm{d}\mu} = 0 = \left(\mu\frac{\mathrm{d}m^2}{\mathrm{d}\mu}\frac{\mathrm{d}V_{\text{tree}}}{\mathrm{d}m^2} + \mu\frac{\mathrm{d}\lambda}{\mathrm{d}\mu}\frac{\mathrm{d}V_{\text{tree}}}{\mathrm{d}\lambda} + \gamma h\frac{\mathrm{d}V_{\text{tree}}}{\mathrm{d}h}\right) + \frac{\mathrm{d}V_{\text{CW}}}{\mathrm{d}h}. \tag{9.22}$$

Indeed, a simple calculation shows that, apart from the cosmological constant, the scale dependence of the tree-level and Coleman–Weinberg potential has the form $\gamma h\frac{\mathrm{d}V_{\text{tree}}}{\mathrm{d}h}$. Specifically,

$$\left(\mu\frac{\mathrm{d}m^2}{\mathrm{d}\mu}\frac{\mathrm{d}V_{\text{tree}}}{\mathrm{d}m^2} + \mu\frac{\mathrm{d}\lambda}{\mathrm{d}\mu}\frac{\mathrm{d}V_{\text{tree}}}{\mathrm{d}\lambda}\right) + \frac{\mathrm{d}V_{\text{CW}}}{\mathrm{d}h} = \frac{3}{64\pi^2}(g_1^2 + 3g_2^2 - 4y_t^2). \tag{9.23}$$

Regardless, the Parwani resummation scheme does not cancel the scale dependence of the quadratic thermal correction,

$$V_T = \left(\frac{3}{32}g_2^2 + \frac{1}{32}g_1^2 + \frac{\lambda}{4} + \frac{y_t^2}{8}\right)T^2x^2 + \cdots. \tag{9.24}$$

The thermal masses for the left- and right-handed fermions have a slightly different form, so let us ignore the top Yukawa as we merely want to show that Parwani resummation does not achieve a cancellation in the scale dependence. The Longtitudinal gauge bosons have a thermal mass $\delta m_Z^2 \sim \frac{11}{6}\frac{g_1^4+g_2^4}{g_1^2+g_2^2}$, $\delta m_W^2 = \frac{11}{6}g_2^2$ and the scalar fields obtain the corrections $\delta m_H^2 = \delta m_G^2 = 2\left(\frac{3}{32}g_2^2 + \frac{1}{32}g_1^2 + \frac{\lambda}{4} + \frac{y_t^2}{8}\right)T^2$. Again, the methods for calculating these corrections we show in chapter 7. It is straightforward to show that the scale dependence resulting from these corrections have the form

$$\begin{aligned}\mu\left(\frac{\mathrm{d}V_T}{\mathrm{d}\mu} + \frac{\mathrm{d}V_{\text{CW}}}{\mathrm{d}\mu}\right) = \frac{T^2}{768\pi^2}\Big(&-81g_2^4 + 3g_1^4 + g_1^2(9g_2^2 - 54\lambda) - 162g_2^2\lambda \\ &+ 144\lambda^2\Big) + \cdots.\end{aligned} \tag{9.25}$$

In principle, one could add all diagrams of $O(g_2^4)$, $O(g_1^4)$, $O(g_1^2g_2^2)$, $O(g_1^2\lambda)$, $O(g_2^2\lambda)$, $O(\lambda^2)$, $O(g_1^2y_t^2)$, $O(g_2^2y_t^2)$, $O(g_3^2y_t^2)$, $O(\lambda y_t^2)$, but this becomes somewhat cumbersome and it is non-trivial to avoid double counting issues. So while the Parwani resummation method handles some of the most dangerous contributions to the breakdown of perturbation theory, it fails to cancel scale dependence up to $O(g^2/\pi^2)$[1] or $O(g^3/\pi^3)$ in the simplest theory and fixing its short

[1] Here and throughout, we are using the usual power counting scheme that the Higgs quartic coupling has the size $\lambda \sim g^2$

comings becomes quickly unmanageable for a more complicated theory. It turns out that the breakdown in perturbation theory, even after the crude resummation of the Parwani method is numerically quite important. For instance, the predicted gravitational wave amplitude predicted using a similar resummation scheme for the simplest extension of the Standard Model varies by multiple orders of magnitude with the value of the scale [3]. The breakdown of perturbation theory is confirmed by the fact that two-loop calculations predict a gravitational wave peak amplitude that is outside the range of uncertainty predicted by the leading order resumed calculation described above. In order to perform an accurate calculation, all contributions must be included to $O(g^4)$. The only known way of performing this (or deriving the potential to be correct to any power in g) is a method known as dimensional reduction. However, the Parwani method does correctly give us the form of corrections at $O(g^3/\pi)$, which we have so far neglected. In the high temperature expansion, they have the form

$$\frac{T}{12\pi}\sum_{i\in\text{bosons}} m_i^3 \to \frac{T}{12\pi}\sum_{i\in\text{bosons}} (m_i^2+\delta m_i^2)^{3/2}. \tag{9.26}$$

The vast majority of the literature has this above term as the sole term from resummation included (although the vast majority of the literature does not include the numerically important wave function renormalization which is needed for the theory to be correct to the same order). The full potential after resummation then has the form,

$$\begin{aligned} V_{1\text{-loop}} = {} & V_{\text{tree}} + V_{\text{CW}}(M^2) + V_T(M^2) - \frac{T}{12\pi}\sum_{i\in\text{bosons}} n_i(m_i^2+\delta m_i^2)^{3/2} \\ & + \frac{T}{12\pi}\sum_{i\in\text{bosons}} n_i m_i^3. \end{aligned} \tag{9.27}$$

Here n_i is the degree of freedom factor. In the authors opinion, it makes little sense not to high temperature expand at least the bosonic thermal functions, as the thermal masses themselves are derived using a high temperature expansion (as will be explained in a later chapter). In any case, the above form of the potential which only includes resumation up to $O(g^3/\pi)$ is known as Arnold–Espinosa resummation [4] and is the most used in the literature. It is not obvious how to include this resummation in a gauge invariant manner outside of dimensional reduction techniques. This means that in practice one has to pick their poison, so to speak, if one wants to do a rough calculation which might be appropriate for phenomological scans of models to obtain a qualitative picture. Either one chooses a gauge independent method that is accurate to $O(g^2)$, or a gauge independent method to $O(g^3/\pi)$, or a significantly more difficult calculation in dimensional reduction at next to leading order.

As mentioned before, there is one more issue we so far have not covered that plagues all resummation methods, even dimensional reduction. Recall Linde's infrared problem that infrared divergences begin to appear when $T/M \to 0$. In the case where the mass is negative, the resumed mass might go to zero for small field values. That is, the thermal corrections cancel the tree- level mass and resummation actually makes this infrared divergence worse and not better. Note that this was disguised by our focus on scale dependence above—for a ϕ^4 theory, Parwani resummation seems to resolve the problems with perturbation theory, however, this infrared divergence is indeed still there if $m^2 < 0$! Further, in dimensional reduction at next to leading order the scale dependence will cancel at $O(g^3)$ but this infrared divergence does not emerge in an obvious way.

A theoretical method of handling this remaining divergence is an open problem and at the moment, the only known solution is a numerical Monte-Carlo simulation. Indeed, it is likely this issue that is behind why both dimensional reduction and Arnold–Espinosa/Parwani resummation techniques give very poor estimates for the critical Higgs mass, below which the electroweak symmetry breaking transition ceases to be first order. Fortunately, the issue may not be terrible for the purposes of electroweak baryogenesis. This paradigm requires not just a first-order transition, but a strong one and any gravitational wave signal that is observable needs a very strong transition. In such a case the mass $m_H^2 + \delta m_H^2$ is never small compared to the temperature so long as we are at or below the critical temperature. So it is expected, but not proven, that perturbation theory is perfectly fine for any transition of phenomelogical interest.

9.1 A brief introduction to dimensional reduction

In the above we introduced the concept of resummation and explained why commonly used methods are inadequate, motivating a more sophisticated approach. Indeed, in figure 9.1 we compare the gravitational wave signal in the commonly used Arnold–Espinosa resummation without RG improvement and wave function renormalization compared to dimensional reduction at next to leading order in the Standard Model effective field theory [4].

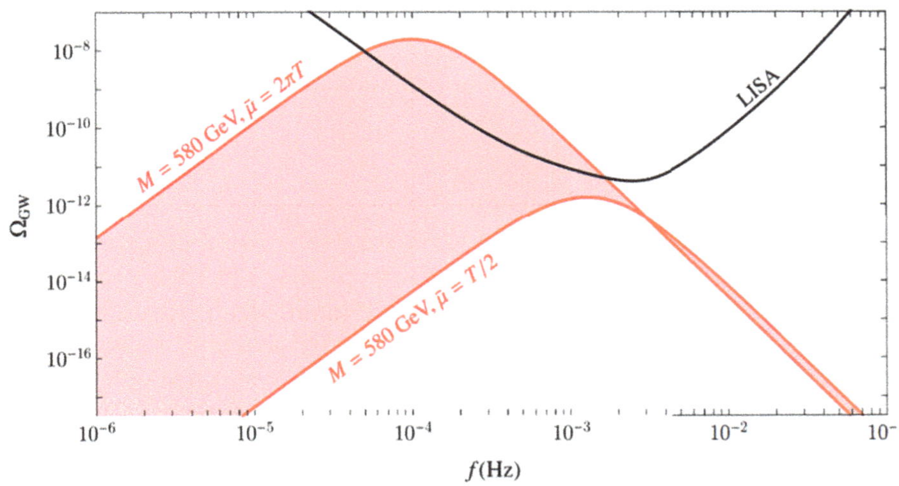

Figure 9.1. Theoretical uncertainty in the gravitational wave spectrum utilizing the most common perturbation theory prescription (including 1 loop finite temperature corrections to the effective potential without RG improvement and resummed daisy diagrams as per the Arnold–Espinosa scheme) for a benchmark in the Standard Model Effective Theory. Figure taken from [3].

Even with RG improvement, the uncertainty is still multiple orders of magnitude. In this chapter we will discuss the best known solution to this problem-dimensional reduction [5]. In doing so, unfortunately, we will have to partially break our promise to be self-contained. A full calculation of the dimensionally reduced Standard Model at the accuracy required to improve on the Parwani or Arnold–Espinosa schemes involves the calculation of around a hundred diagrams at finite temperature. Theories beyond the Standard Model require the calculation of many more diagrams, though in principle the calculation can be automated. In this chapter, we will have to content ourselves to equipping the reader with the skills needed to perform a more detailed calculation.

Dimensional reduction begins by noticing that propagators at finite temperature in the imaginary time formalism has the structure you would expect from a three-dimensional Kaluza–Klein theory. In particular, for bosons and fermions respectively, we have

$$G = \frac{1}{\omega_n^2 + k^2 + m^2}, \qquad S = \frac{\gamma^0\omega_n + \vec{\gamma}\bullet\vec{k} + m}{\omega_n^2 + k^2 + m^2}, \tag{9.28}$$

where $\omega_n = 2n\pi T$ and $\omega_n = (2n + 1)\pi T$ for fermions and bosons, respectively, as in chapter 8. This is the propagator one would have if you had a three-dimensional theory, with the fourth compactified dimension having a length $1/T$. The Matsubara frequencies can be interpreted as higher Kaluza–Klein modes with an effective mass $\sqrt{\omega_n^2 + m^2}$. This means we can integrate out heavy Kaluza–Klein modes that gives a three-dimensional effective field theory that is valid at scales below the lightest Matsubara mode, $p<\pi T$.

The kinetic part of the Standard Model gauge fixing Lagrangian is as follows,

$$L \sim \frac{1}{4}G^a_{\mu\nu}G^{a,\,\mu\nu} + \frac{1}{2\xi}(\partial_\mu A^a_\mu)^2 + \sum_i \overline{\psi}_i\gamma^\mu D_\mu\psi_i + \partial_\mu\bar{\eta}^a D_\mu\eta^a + |\, D_\mu\phi\,|^2 + \cdots. \tag{9.29}$$

with an effective potential

$$V = \frac{1}{2}m^2\phi^2 + \frac{1}{4}\lambda \quad \phi^4. \tag{9.30}$$

In other words, we are considering the Standard Model without hypercharge for simplicity. After integrating out all Matsubara modes, the theory will become independent of fermions apart from the matching conditions. Further, the temporal component of the gauge bosons will acquire a thermal mass, a thermally induced quartic self-interaction and ultimately act as a scalar. The effective potential in our 3D dimensionally reduced effective field theory has the form

$$\begin{aligned} V_{3D} = m_3^2\,|\phi|^2 + \lambda\,|\phi|^4 + \frac{1}{2}m_D^2 A_0^a A_0^a + \frac{1}{2}m_{D,\,G}^2 G_0^a G_0^a + \frac{1}{4}\kappa_1(A_0^a A_0^a)^2 \\ + h_1\,|\phi|^2 A_0^a A_0^a + h_2\,|\phi|^2 G_0^a G_0^a. \end{aligned} \tag{9.31}$$

We can equate effective vertices in both the 4 and 3D theories and match in order to acquire the form of the above couplings. The master formula governing the relationship between three- and four-dimensional fields from which one can acquire the above parameters is

$$\varphi_{3D}^2 = \frac{1}{T}\varphi_{4D}^2(1 + \hat{\Pi}_{\varphi}'), \tag{9.32}$$

where φ is a generic scalar field in our theory and $\hat{\Pi}_{\varphi}'$ is the momentum derivative of the renormalized self-energy—that is, counter terms are included. The fact that we consistently include the self-energies at finite temperature as well as the counter terms means we automatically perform a resummation and the correction from the wave function renormalization is automatically taken into account. Let us give an illustrative example for matching, if we wish to match a quartic interaction

$$T(-2\lambda_3)\phi_3^4 = (-2\lambda + \hat{\Gamma}_{\phi^4})\phi^4, \tag{9.33}$$

where we have cancelled the loop correction to the four-point function of the three-dimensional theory, which is equal to the soft contribution to the four-point function in the four-dimensional theory. Substituting equation (9.32) into the above we then get the matching relationship for the quartic

$$T\lambda_3 = T\left(\lambda - \frac{1}{2}\hat{\Gamma}_{\phi^4}\right)\phi^4. \tag{9.34}$$

On the other hand, matching the two-point function is a straightforward application of equation (9.32)

$$m_3^2 = m^2 + \Pi_{\phi^2}. \tag{9.35}$$

It is quite a tedious task to calculate all diagrams in the dimensionally reduced potential. For brevity we will just discuss the one-loop contributions the m_D^2 mass term in the Landau gauge and refer the reader to references [3, 5–7] for more details. To calculate m_D^2 we need the one-loop two-point function.

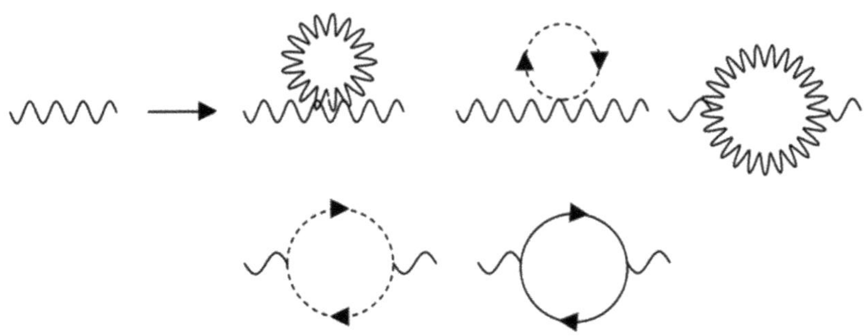

We will also neglect scalars in the loops as this is fairly similar to the diagram calculated in the previous chapter. We need the Feynman rules for the gauge bosons self-interactions in order to proceed. These are

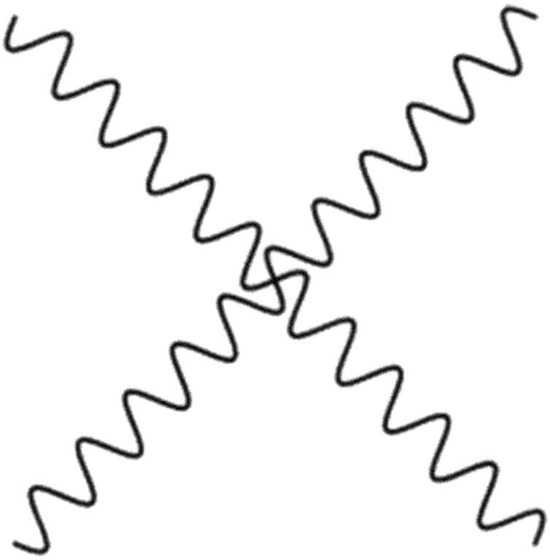

$$
\begin{aligned}
&= g^2\Delta^{\mu\nu\kappa\lambda}_{abcd} \\
&= g^2[\delta_{ab}\delta_{cd}(\delta_{\mu\kappa}\delta_{\nu\lambda} + \delta_{\mu\lambda}\delta_{\nu\kappa} - 2\delta_{\mu\nu}\delta_{\kappa\lambda}) + \delta_{ac}\delta_{bd}(\delta_{\mu\nu}\delta_{\kappa\lambda} + \delta_{\mu\lambda}\delta_{\nu\kappa} - 2\delta_{\mu\kappa}\delta_{\nu\lambda}) \\
&\quad + \delta_{ad}\delta_{bc}(\delta_{\mu\nu}\delta_{\kappa\lambda} + \delta_{\mu\kappa}\delta_{\nu\lambda} - 2\delta_{\mu\lambda}\delta_{\nu\kappa})]
\end{aligned} \tag{9.36}
$$

Beginning with the bubble diagram we have

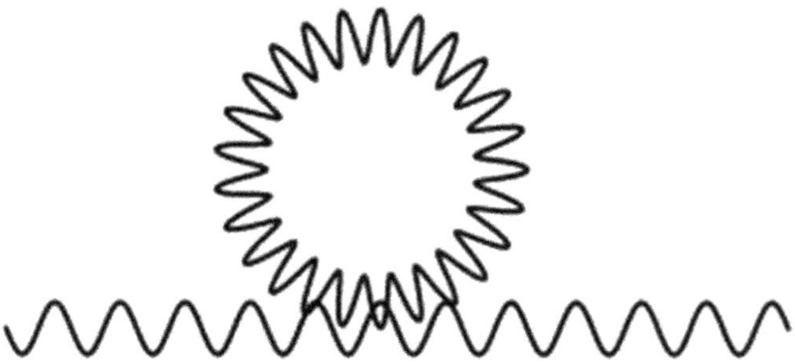

$$
\begin{aligned}
&= \frac{1}{2}g^2T\left(\frac{\mu^2e^{\gamma_E}}{4\pi}\right)\sum_{\omega_n\neq 0}\int\frac{d^dP}{(2\pi)^d}\Delta^{\mu\nu\kappa\lambda}_{abcd}\delta_{cd}\frac{1}{P^2}\left(\delta_{\kappa\lambda} - \frac{P_\kappa P_\lambda}{P^2}\right) = 2g^2\delta_{ab}T\left(\frac{\mu^2e^{\gamma_E}}{4\pi}\right) \\
&\sum_{\omega_n\neq 0}\int\frac{d^dP}{(2\pi)^d}\left(\frac{(2-D)}{P^2}\delta_{\mu\nu} - \frac{P_\mu P_\nu}{P^4}\right).
\end{aligned} \tag{9.37}
$$

A simple calculation gives

$$g^2\delta_{ab}I_1^{4b}\begin{cases} -(D-1), & \mu=\nu=0 \\ -(-1+2(D-1)), & \mu=i,\, \nu=j \end{cases}, \tag{9.38}$$

where the master integral has the form

$$I_1^{4b} = \frac{T^2}{12}\left(\frac{\mu}{4\pi T}\right)^{2\epsilon}\left\{1+2\left[\log[2\pi]+\gamma-\frac{\zeta'(2)}{\zeta(2)}\right]\varepsilon\right\}. \tag{9.39}$$

This piece is finite, however other terms in the figure contain a pole such as the fermion loop.

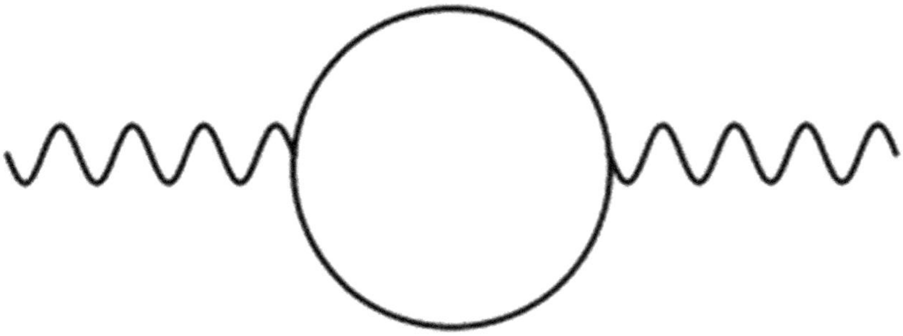

$$= N_f(1+N_c)\left(\frac{i}{2}g\right)^2(\tau_a)^{im}(\tau_b)^{nj}\delta_{ij}\delta_{mn}\left(\frac{\mu^2 e^{\gamma_E}}{4\pi}\right)$$

$$\sum_{\omega_n}\int\frac{d^dQ}{(2\pi)^4}\frac{1}{Q^2(Q+K)^2}Tr\left[P_L iQ^\alpha\gamma_\alpha\gamma_\mu P_L i(Q+K)^\beta\gamma_\beta\gamma_\nu\right].$$

To quadratic order in K, this simplifies to

$$g^2N_f(1+N_c)\delta_{ab}\begin{cases}(2^{2-d}-1)(d-1)I_1^{4b}+\frac{K^2}{6}(2^{4-d}-1)(d-1)I_2^{4b}, & (\mu=\nu=0) \\ -\frac{1}{3}(2^{4-d}-1)I_2^{4b}(K^2\delta_{mn}-K_mK_n), & (\mu=m, \quad \nu=n)\end{cases}.$$

In this case the master integral has a pole and is of the form

$$I_2^{4b} = \frac{1}{16\pi^2}\left(\frac{\Lambda}{4\pi T}\right)^{2\epsilon}\left[\frac{1}{\epsilon}+2\gamma\right].$$

Finally, we have

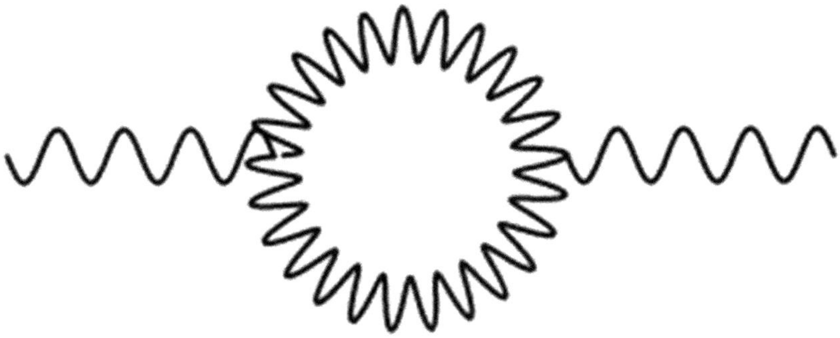

$$(-ig)2 = \left(\frac{\mu^2 e^{\gamma_E}}{4\pi}\right)\sum_{\omega_n\neq 0}\int \frac{\mathrm{d}^d P}{(2\pi)^d}\Delta^{\mu\sigma\lambda}_{abc}(K,\, P,\, -Q)\Delta^{\nu\alpha\rho}_{bef}(-K,\, -P,\, Q)$$
$$\delta_{cf}\frac{1}{Q^2}\left(\delta_{\lambda\rho} - \frac{Q_\lambda Q_\rho}{Q^2}\right)\delta_{de}\frac{1}{P^2}\left(\delta_{\sigma\alpha} - \frac{P_\sigma P_\alpha}{P^2}\right).$$

For this we need the Feynmann rule for the tripple gauge coupling,

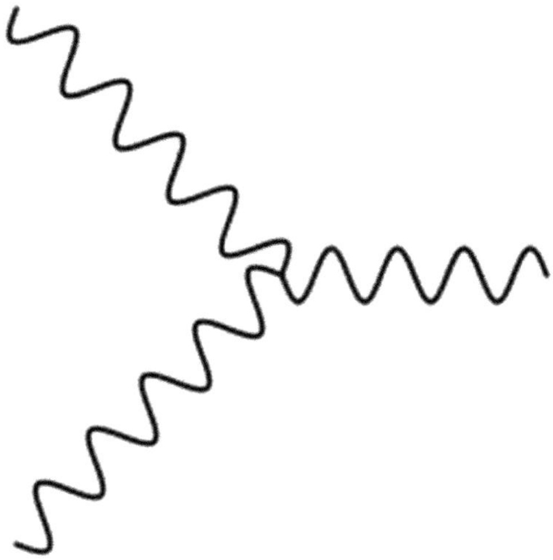

$$= \Delta^{\mu\nu\lambda}_{abc}(K,\, P,\, Q) = -ig\epsilon^{abc}((P - Q)_\mu\delta_{\nu\lambda} + (Q - K)_\nu\delta_{\lambda\mu} + (K - P)_\lambda\delta_{\mu\nu})$$

Again, we can expand in K to quadratic order to obtain in the Landau gauge,

$$g^2\delta_{ab}\begin{cases} d(4-2d)I_1^{4b} + \frac{K^2}{6}(16-3d + 2d^2)I_2^{4b} \quad (\mu = \nu = 0) \\ 2d\ \delta_{mn}I_1^{4b} + \frac{1}{6}(K^2\delta_{mn}(31-2d) - 2K_mK_n(17 - d)) \quad (\mu = m, \quad \nu = n) \end{cases}.$$

After cancelling poles with the appropriate counter terms and replacing $K = gT$, one finds that

$$m_D^2 = \hat{\Pi}_{A_0^a A_0^b} = g^2T^2\left(\frac{5}{6} + \frac{N_f}{3}\right). \tag{9.40}$$

For more details as well as the calculation of the two-loop contributions involving around 100 diagrams calculated at finite temperature, the reader is referred to references [3, 5].

The final step in dimensional reduction is to realize that the dimensionally reduced theory still has a heirachy. The temporal gauge bosons are at a scale $m_D \sim gT$ and the mass scale or the resummed scalar is typically lower than this. In this case we integrate out the gauge boson scale, known as a 'soft scale' to write the effective potential at the so called ultrasoft scale

$$V_{3D}^{\text{ultrasoft}} = \overline{m}_3^2\phi^2 + \overline{\lambda}_3\phi^4. \tag{9.41}$$

The full solution in the Standard model is quite gory, as one only achieves dramatic improvement over conventional resummation techniques when one includes two-loop effects. For self-contained derivations of the Standard Model, the Standard Model Effective field theory and the singlet extension of the Standard Model, see references [3, 5, 6].

In the above we gave a sketch of the method involved in dimensional reduction while referring to other papers for the details for any realistic model. However, it is necessary to demonstrate the utility of the approach—it allows one to correctly define the potential to the correct power in the coupling. One only dramatically improves upon conventional resummation techniques shown in the first part of this chapter if a two-loop calculation involving around 100 diagrams for the simplest model is performed. This makes the demonstration of the improvement difficult, even for a textbook of this length. As a compromise we will give the two-loop effective potential of the ϕ^4 theory with a four-dimensional effective potential as before

$$V_{\text{tree}} = \frac{1}{2}m^2\varphi^2 + \frac{g^2}{4!}\varphi^4. \tag{9.42}$$

and demonstrate that it succesfully improves upon conventional resummation methods. The thermal potential is [7]

$$\begin{aligned} V = T\Bigg(&\frac{1}{2}m_3^2v_3^2 + \frac{1}{24}g_3^2v_3^4 - \frac{(M_3^2)^{3/2}}{12\pi} \\ &+ \frac{1}{16\pi^2}\left[\frac{1}{8}g_3^2M_3^2 - \frac{1}{24}g_3^4v_3^2\left(1+2\log\left[\frac{\mu_3}{2M_3}\right]\right)\right]\Bigg). \end{aligned} \tag{9.43}$$

Here, μ_3 is the renormalization scale of the dimensionally reduced theory and the matching relations are

$$v_3 = \frac{\phi}{\sqrt{T}}, \tag{9.44}$$

$$g_3^2 = T\left(g^2 - \frac{3}{32\pi^2}g^4L\right), \tag{9.45}$$

$$m_3^2 = m^2 + \frac{1}{24}g^2T^2 - \frac{1}{(4\pi)^2}\left(\frac{1}{2}g^2m^2L + \frac{1}{16}g^4T^2L + \frac{1}{6}g_3^4\left(c + \log\left[\frac{3T}{\mu_3}\right]\right)\right), \tag{9.46}$$

$$M_3^2 = m_3^2 + \frac{1}{2}g_3^2v_3^2. \tag{9.47}$$

To demonstrate the cancellation of scale dependence in the effective potential, it is easiest to proceed by considering the scale dependence of each of the above parameters in the effective theory. Using the fact that $\mu\frac{\mathrm{d}L_b}{\mathrm{d}\mu} = 2$, it is easy to show that

$$\mu\frac{\mathrm{d}g_3^2}{\mathrm{d}\mu} = O(g^6/\pi^4), \quad \mu\frac{\mathrm{d}m_3^2}{\mathrm{d}\mu} = \mu\frac{\mathrm{d}M_3^2}{\mathrm{d}\mu} = O(g^4/\pi^4), \tag{9.48}$$

and

$$\mu_3\frac{\mathrm{d}m_3^2}{\mathrm{d}\mu_3} = \frac{1}{96\pi^2}g_3^2. \tag{9.49}$$

This fascillitates the leading order cancellation of

$$\mu_3\frac{\mathrm{d}}{\mathrm{d}\mu_3}\left(\frac{1}{2}m_3^2v_3^2 - \frac{1}{4\pi^2}\left[\frac{1}{24}g_3^4\left(1{+}2\log\left[\frac{\mu_3}{2M_3}\right]\right)\right]\right) \sim 0. \tag{9.50}$$

where the equation becomes an equality if one ignores the scale dependence of M_3 in the log. The leading order piece of scale dependence which is uncancelled is

$$\mu_3\frac{\mathrm{d}}{\mathrm{d}\mu_3}\left(\frac{(M_3^2)^{3/2}}{12\pi}\right) \sim \frac{1}{12\pi}\frac{3}{2}\frac{1}{96\pi^2}T^3g^4M_3. \tag{9.51}$$

The leading order field dependent term adds a further power of g to this, showing that the non cancellation as of $O(g^5/\pi^3)$. This is a substantial improvement on the resummation methods of Parwani or Arnold–Espinosa, where the cancellation of terms of $O(g^4/\pi^2)$ required the inclusion by hand of lollipop diagrams and no cancellation existed at all for terms in the effective cubic of $O(g^3/\pi^3)$.

References

[1] Linde A 1980 Infrared problem in thermodynamics of the Yang–Mills gas *Phys. Lett.* B **96** 289–92

[2] Parwani R 1992 Resummation in a hot scalar field theory *Phys. Rev.* D **45** 4695

[3] Croon D and Gould O; Schicho P, Tenkanen T and White G 2021 Theoretical uncertainties for cosmological first order phase transitions *J. High Energy Phys.* **2021** 055

[4] Arnold P and Espinosa O 1993 The effective potential and first order phase transitions: beyond leading-order *Phys. Rev.* D **47** 3546

[5] Kajantie K, Laine M, Rummukainen K and Shaposhnikov M 1994 Generic rules for high temperature dimensional reduction and their application to the standard model *Nucl. Phys.* B **458** 90

[6] Brauner T, Tenkanen T, Tranberg A, Vuorinen A and Weir D 2017 Dimensional reduction of the Standard Model coupled to a new singlet scalar field *J. High Energy Phys.* **03** 007

[7] Gould O and Tenkanen T 2021 On the perturbative expansion at high temperature and implications for cosmological phase transitions *J. High Energy Phys.* **06** 069

IOP Publishing

Electroweak Baryogenesis (Second Edition)
An introduction
Graham White

Chapter 10

Transport equations

We now have enough information to build a network of transport equations for a given model during a phase transition. In chapter 6 we derived how to relate the divergences of particle number current densities to functions of the particle's self-energies. In the previous chapter we derived the thermal masses and thermal widths which are inputs for the transport coefficients and CP violating sources. We also derived diffusion coefficients and bubble wall velocities as they will be useful in relating the divergence of the current density to derivatives in a single space–time variable of the number densities. In this section we finally put this all together in the MSSM. As we will discuss in chapter 14, electroweak baryogenesis in the minimal supersymmetric Standard Model (MSSM) is either dead or at least on life support. Even still, the MSSM contains scalar and fermionic sources as well as the knowledge of tricks used in many other models. Thus, the MSSM retains a status as being the most useful test model for pedagogical purposes.

10.1 The MSSM under supergauge equilibrium

A full calculation of charge transport during the electroweak transition requires solving a set of coupled differential equations with one equation per particle species (treating left doublets and right handed particles as separate species). We can immediately make a simplification by assuming only the top Yukawa has interactions fast enough amongst fermions to be numerically important. Second, we can assume that gauge and supergauge interactions are fast enough to allow us to consider the number densities of gauge singlets, summed with their superpartners as a single number density. Furthermore, there are some fast interactions that allow us to assume local equilibrium between the two Higgs doublets. Then, our network of transport equations only involve three number densities

doi:10.1088/978-0-7503-3571-3ch10

$$\begin{aligned} Q &= n_{t_L} + n_{b_L} + n_{\tilde{t}_L} + n_{\tilde{b}_L} \\ T &= n_{t_R} + n_{\tilde{t}_R} \\ H &= n_{H_u^+} + n_{H_u^0} - n_{H_d^-} - n_{H_d^0} n_{\tilde{H}_u^+} + n_{\tilde{H}_u^0} - n_{\tilde{H}_d^-} - n_{\tilde{H}_d^0}. \end{aligned} \tag{10.1}$$

We can relate chemical potentials to number densities using an assumption of local thermal equilibrium. Ignoring terms of $O(\mu^3/T^3)$, that is, performing a high temperature expansion, we can derive the relation

$$\mu_x = \frac{6}{T^2}\frac{n_x}{k_x} \tag{10.2}$$

with

$$k_x = k_x(0)\frac{c_{F,\,B}}{\pi^2}\int_{m/T}^{\infty} \mathrm{d}yy\frac{\mathrm{e}^y}{(\mathrm{e}^y \pm 1)^2}\sqrt{y^2 - m_x^2/T^2}, \tag{10.3}$$

where $c_{F(B)} = 6(3)$ and the sign in the denominator is $\pm$ for fermions and bosons, respectively. The factors $k_i(0)$ are 2 for Dirac fermions and complex scalars and 1 for chiral fermions. The k factors of our composite charge densities in equation (10.3) are the sum of the k factors for each component. Combining our number densities together in this way means we also combine the vacuum expectation value (VEV) insertion diagrams of tops and stops as well as Higgs and Higgsinos to each form a singlet transport coefficient:

$$\Gamma_M^{\pm} = \Gamma_M^{\pm}\tilde{t} + \Gamma_M^{\pm}t\Gamma_H^{\pm} = \Gamma_M^{\pm}\tilde{H}^{\pm} + \Gamma_M^{\pm}\tilde{H}^0. \tag{10.4}$$

Furthermore, all one-loop three-body rates involving t, Q and H and their superpartners, as cataloged in chapter 6, sum into a single three-body rate, Γ_Y, for the linear combinations of composite densities $\mu_t - \mu_H - \mu_Q$. The set of coupled differential equations is

$$\begin{aligned} \partial_\mu T^\mu = {} & \Gamma_M^+\left(\frac{T}{k_T} + \frac{Q}{k_Q}\right) - \Gamma_M^-\left(\frac{T}{k_T} - \frac{Q}{k_Q}\right) - \Gamma_Y\left(\frac{T}{k_T} - \frac{H}{k_H} - \frac{Q}{k_Q}\right) \\ & + \Gamma_{SS}\left(\frac{2Q}{k_Q} - \frac{T}{k_T} + \frac{9(Q+T)}{k_B}\right) + S_{\tilde{t}}^{C/P} \end{aligned} \tag{10.5}$$

$$\begin{aligned} \partial_\mu Q^\mu = {} & -\Gamma_M^+\left(\frac{T}{k_T} + \frac{Q}{k_Q}\right) - \Gamma_M^+\left(\frac{T}{k_T} - \frac{Q}{k_Q}\right) + \Gamma_Y\left(\frac{T}{k_T} - \frac{H}{k_H} - \frac{Q}{k_Q}\right) \\ & -2\Gamma_{SS}\left(\frac{2Q}{k_Q} - \frac{T}{k_T} + \frac{9(Q+T)}{k_B}\right) - S_t^{C/P} \end{aligned} \tag{10.6}$$

$$\partial_\mu H^\mu = -\Gamma_H\frac{H}{k_H} + \Gamma_Y\left(\frac{T}{k_T} - \frac{Q}{k_Q} - \frac{H}{k_H}\right) + S_{\tilde{H}}^{C/P}, \tag{10.7}$$

where the strong sphaleron rate is taken to be $\Gamma_{SS} \approx 16\alpha_S^4 T$ [1].

10.2 Solution using fast rates, diffusion approximation, and ultrathin wall approximations

Now we have a network of coupled transport equations we will outline the simplest approach to solving them. Unfortunately, the approach produces a number for the baryon asymmetry that disagrees with more complete methods by up to two orders of magnitude. Nonetheless, it is useful for pedagogical purposes, but we will not go to lengths to justify the approximation we use as they can be viewed as a stepping stone to more complete methods.

From the solution to the network of coupled transport, equations manifest particle–anti-particle asymmetries that diffuse into the symmetric phase ahead of the bubble wall, biasing electroweak sphalerons. The expanding bubble wall will capture some of the net baryon asymmetry. The first step in finding an approximate solution to the network of transport equations is to convert the left-hand sides of the network of transport equations into derivatives of number densities in a single space–time variable. This is achieved by first using the diffusion approximation that we introduced earlier

$$\partial_\mu J^\mu = v_W \dot{n} - D_n \nabla^2 n, \tag{10.8}$$

then making an assumption that the bubble wall is symmetric and we can ignore the curvature of the bubble wall, so that we can reduce the problem to a one-dimensional one. Shifting to the bubble wall rest frame using the variable $z \equiv | v_w t - x |$, we can write the set of coupled transport equations as a set of differential equations in z, whose left-hand sides look like $v_w n_x' - D_x n_x''$. Next we assume both Γ_Y and Γ_{SS} are fast enough such that the linear combination of number densities that these rates are coefficients for are in local equilibrium. That is,

$$\begin{aligned} \frac{T}{k_T} - \frac{H}{k_H} - \frac{Q}{k_Q} &= 0 \\ \frac{2Q}{k_Q} - \frac{T}{k_T} + \frac{9(Q+T)}{k_B} &= 0. \end{aligned} \tag{10.9}$$

The second equation can be used to solve T in terms of Q and both can be solved in terms of H using the first equation. The result is,

$$\begin{aligned} Q &= \frac{(k_B - 9k_T)k_Q}{(9k_T + 9k_Q + k_B)k_H} H \\ T &= \frac{(9k_T + 2k_B)k_T}{(9k_T + 9k_Q + k_B)k_H}. \end{aligned} \tag{10.10}$$

We have reduced our set of transport equations down to a single one. The remaining linearly independent transport equation is any that does not have any fast rate on the right-hand side. One such combination is 2 × (10.5) + (10.6) + (10.7). In this case the transport equation is

$$v_w H'(z) - \bar{D} H''(z) + \bar{\Gamma} H(z) = S(z), \tag{10.11}$$

where

$$\bar{D} = \frac{(9k_Q k_T + k_B k_Q + 4k_T k_B)D_q + k_H(9k_T + 9k_Q + k_B)D_h}{9k_Q k_T + k_B k_Q + 4k_T k_B + k_H(9k_Q + 9k_T + k_B)}$$
$$\bar{\Gamma} = \frac{(9k_Q + 9k_T + k_B)(\Gamma_M^- + \Gamma_H) - ((3k_B + 9k_Q - 9k_T)\Gamma_M^+)}{9k_Q k_T + k_B k_Q + 4k_T k_B + k_H(9k_Q + 9k_T + k_B)} \tag{10.12}$$
$$S(z) = \frac{k_H(9k_Q + 9k_T + k_B)}{9k_T k_Q + k_B k_Q + 4k_T k_B + k_H(9k_Q + 9k_T + k_B)}\left(S_{\bar{t}} + S_{\bar{H}}\right).$$

Next we make the ultrathin wall approximation where the Higgs profile is assumed to be close to a step function, meaning $S(z)$ and $\bar{\Gamma}$ are zero in the symmetric phase, i.e. when $z > 0$. Furthermore, the functions $\bar{\Gamma}$ become constant in the broken phase in this approximation. This linearizes the remaining transport equations and allows us to solve it in two regions using the method of variable coefficients. The solution in the symmetric phase, up to yet-to-be-determined integration constants that are specified by the boundary and matching conditions, is

$$H(z) = A e^{v_w z/\bar{D}} + C \tag{10.13}$$

and the solution in the broken phase is

$$H(z) = \frac{e^{\kappa_+ z}}{\bar{D}(\kappa_+ - \kappa_-)}\left(\int_0^z dy e^{-\kappa_+} S(y) - \beta_+\right)$$
$$-\frac{e^{\kappa_- z}}{\bar{D}(\kappa_+ - \kappa_-)}\left(\int_0^z dy e^{-\kappa_-} S(y) - \beta_-\right). \tag{10.14}$$

Here the exponentials solve the homogeneous equations and the exponents have the form

$$\kappa_\pm = \frac{v_w \pm \sqrt{v_w^2 + 4\bar{\Gamma}\bar{D}}}{2\bar{D}} \tag{10.15}$$

We have two boundary conditions in that the densities must go to zero at $\pm\infty$ which sets C to zero and β_+ to

$$\beta_+ = \int_0^\infty dy e^{-\kappa_+ y} S(y). \tag{10.16}$$

The other two integration constants are determined by the fact that both $H(z)$ and $H'(z)$ are both continuous at the boundary, $z = 0$. The matching conditions are then

$$A = \frac{\beta_- - \beta_+}{\bar{D}(\kappa_+ - \kappa_-)}$$
$$A(\kappa_+ + \kappa_-) = \frac{\beta_- \kappa_- - \beta_+ \kappa_+}{\bar{D}(\kappa_+ - \kappa_-)}. \tag{10.17}$$

Solving the above equations gives

$$\begin{aligned} \beta_- &= \beta_+ \frac{\kappa_-}{\kappa_+} = \frac{\kappa_-}{\kappa_+} \int_0^\infty \mathrm{d}y \mathrm{e}^{-\kappa_+ y} S(y) \\ A &= -\frac{\beta_+}{\kappa_+} = \frac{1}{\bar{D}\kappa_+} \int_0^\infty \mathrm{d}y \mathrm{e}^{-\kappa_+ y} S(y). \end{aligned} \tag{10.18}$$

10.3 Solution without fast rates

Here we will now resolve the transport equations without making the assumption that the rates are fast. If we otherwise use the same assumptions as before we once again have a set of linearized differential equations which gives us hope for a semi-analytic solution. Note the structure of the transport equations—if we take the combination $\partial_\mu (T + Q)^\mu$ we have a transport equation that depends only on two number densities. If we can create another linear combination that is a function of three number densities only, then the first equation can solve T in terms of Q, the second can solve for H in terms of Q, and the final will have the CP violating source. To create a linear combination of transport equations that is a function of number densities T, Q and H only, we note that the two CP violating sources are proportional to each other

$$S_{\tilde{t}}^{C/P} = \frac{1}{2a} S_{\tilde{H}}^{C/P} \tag{10.19}$$

The set of equations we want is therefore the linear combinations (10.5) + (10.6), (1 $+a$) × (10.5) +(1−a) × (10.6) + (10.7) and 2 × (10.5) + (10.6) + (10.7).

10.4 Deriving the analytic solution

The problem in deriving an analytic solution quickly becomes messy due to all the k factors and transport coefficients. It is a great simplification to rewrite all transport coefficients in the form $a_{Xj}^{\ i}$, where $j \in \{1,2,3\}$ is the equation number, $X \in \{Q,T,H\}$ is the field index, and $i \in \{0,1,2\}$ is the power of the derivative in z

$$a_{Q1}^i \partial^i Q + a_{T1}^i \partial^i T = 0 \tag{10.20}$$

$$a_{Q2}^i \partial^i Q + a_{T2}^i \partial^i T + a_{H2}^i \partial^i H = 0 \tag{10.21}$$

$$a_{Q3}^i \partial^i Q + a_{T3}^i \partial^i T + a_{H3}^i \partial^i H = \Delta(z). \tag{10.22}$$

Here of we have used Einstein's summation convention. To solve the first equation for T in terms of Q, use the method of variable coefficients treating Q as an inhomogeneous source. Doing this gives

$$T = \frac{1}{a_{T1}^2} \sum_{\pm} \frac{1}{\kappa_\mp - \kappa_\pm} \mathrm{e}^{\kappa_\pm z} \left[\int^z \mathrm{e}^{-\kappa_\pm y} \left(a_{Q1}^i \frac{\partial^i Q}{\partial y^i} \right) \mathrm{d}y - \beta_i \right]. \tag{10.23}$$

It was demonstrated in [2] that the integration constants are zero. We will ignore them as they clutter notation and little is learned more than the overall counting exercise—we have three number densities and we take their second derivatives so at most we can have six boundary conditions and six matching conditions which specify a total of twelve integration constants. If we substitute the above solution back into the subsequent transport equations we are left with an integro-differential equation. To overcome this we will make a series of variable changes. First,

$$h_\pm = \int^z e^{-\kappa_\pm y} Q \mathrm{d}y. \tag{10.24}$$

This eliminates the exponent and integral in equation (10.23), but we now have two functions related via the identity

$$h'_+ = e^{(\kappa_- - \kappa_+)z} h'_-. \tag{10.25}$$

Let us make another change of variables to remove the exponential outside the integral

$$j_\pm = e^{\kappa_\pm z} h_\pm. \tag{10.26}$$

Finally, we need to write everything in terms of a single variable. This is achieved via the substitution

$$k = e^{\kappa_\mp z} \int^z e^{-\kappa_\mp y} j_\pm \mathrm{d}y, \tag{10.27}$$

which allows us to relate $j_\pm$ to k with the equation

$$j_\pm = k' - \kappa_\mp k. \tag{10.28}$$

We can then relate both T and Q to derivatives in k, which leaves us with differential equations only. It is convenient to rescale k in a way such that

$$T = -a^i_{Q1} \partial^i k \tag{10.29}$$

$$Q = a^i_{T1} \partial^i k. \tag{10.30}$$

We have achieved our initial aim of rewriting the solutions to the differential equation without awkward derivatives or integrals. One can verify by direct substitution that the above indeed satisfies the first transport equation. Next we substitute these equations for T and Q in terms of k into equation (10.21), which gives a differential equation in terms of k and H only,

$$0 = a^i_{Q2} \partial^i Q + a^i_{T2} \partial^i T + a^i_{H2} \partial^i H \tag{10.31}$$

$$= \left(a^i_{Q2} a^j_{T1} - a^i_{T2} a^j_{Q1} \right) \partial^{i+j} k + a^i_{H2} \partial^i H. \tag{10.32}$$

Using the same method as before one finds that the solution for H is given by

$$H = \frac{1}{a^2_{H2}} \sum_\pm \frac{e^{\kappa_\pm z}}{\kappa_\mp - \kappa_\pm} \int^z e^{-\kappa_\pm y} \left(a^i_{Q2} a^j_{T1} - a^i_{T2} a^j_{Q1} \frac{\partial^{i+j} k}{\partial y^i} \right) \mathrm{d}y. \tag{10.33}$$

Following the same recipe as before we can write H and k in terms of a variable l

$$H = -\sum_{n=0}^{4}\delta_{i+j-n}(a_{Q2}^{i}a_{T1}^{j} - a_{T2}^{i}a_{Q1}^{j})\partial^{n}l \tag{10.34}$$

$$k = a_{H2}^{i}\partial^{i}l. \tag{10.35}$$

In the above we added a Kronecker delta so that the structure of the equations was made more clear. We can then substitute our solution for k into our solutions for T and Q to give all three number densities in terms of l

$$\begin{aligned} H &= -\sum_{n=0}^{4}\delta_{i+j-n}(a_{Q2}^{i}a_{T1}^{j} - a_{T2}^{i}a_{Q1}^{j})\partial^{n}l \\ T &= -\sum_{n=0}^{4}\delta_{i+j-n}a_{Q1}^{i}a_{H2}^{j}\partial^{n}l \\ Q &= \sum_{n=0}^{4}\delta_{i+j-n}a_{T1}^{i}a_{H2}^{j}\partial^{n}l. \end{aligned} \tag{10.36}$$

We now can write equation (10.22) in terms of a single variable, l, and a source

$$\begin{aligned} \Delta(z) &= \sum_{n=0}^{6}\delta_{i+j+k-n}\left(a_{T1}^{i}a_{H2}^{j}a_{Q3}^{k} - a_{Q1}^{i}a_{H2}^{j}a_{T3}^{k}\right)\left(-a_{T1}^{i}a_{Q2}^{j}a_{H3}^{k} + a_{Q1}^{i}a_{T2}^{j}a_{H3}^{k}\right)\partial^{n}l = \\ &\sum_{n=0}^{6}\delta_{i+j+k-n}\epsilon^{abc}a_{Ta}^{i}a_{Hb}^{j}a_{Qc}^{k}\partial^{n}l \equiv \sum_{n=0}^{6}a_{l}^{n}\partial^{n}l. \end{aligned} \tag{10.37}$$

The step that introduced the permutation symbol makes use of the fact that $a_{H1}^{i} = 0$. Once again, using the method of variable coefficients the solution to the above equation is

$$l = \sum_{i=1}^{6}x_{i}e^{\alpha_{i}z}\left(\int^{z} e^{-\alpha_{i}y}\Delta(y)dy - \beta_{i}\right) \tag{10.38}$$

in the EWSB phase and

$$l = \sum_{i=1}^{6}y_{i}e^{\gamma_{i}z} \tag{10.39}$$

in the symmetric phase. In the above we have defined α_i and γ_i as the roots of the polynomials

$$\begin{aligned} \sum_{n=0}^{6}a_{l}^{n}\alpha^{n} &= 0 \\ \sum_{n=0}^{6}a_{l}^{n}\gamma^{n} &= 0, \end{aligned} \tag{10.40}$$

which are the characteristic polynomials in the broken and symmetric phases, respectively. Furthermore the constants x_i are derived from the equation

$$\vec{x} = M^{-1}\vec{d}. \tag{10.41}$$

In the above the matrix $M_{ij} \equiv \alpha_i^{j-1}$ and j is both an exponent and an index ranging from 1 to 6. Also $\vec{d} \equiv [0, \ldots, 1/a_l^6]^{\mathrm{T}}$ The integration constants are denoted by β_i and y_i in the broken and symmetric phases, respectively. They are determined by matching and boundary conditions. The boundary conditions are that all number densities must be null at $\pm\infty$, whereas the matching conditions are that all number densities and their derivatives must be continuous at the bubble wall. For the symmetric phase the boundary conditions are simple to enforce,

$$y_i = 0 \;\; \forall\, \gamma_i \leqslant 0. \tag{10.42}$$

For the broken phase we obtain a condition on the positive exponents

$$x_i\beta_i = x_i \int_0^\infty \mathrm{d}y e^{-\alpha_i y}\Delta(y) \equiv I_i \;\; \forall\, \alpha_i \geqslant 0. \tag{10.43}$$

Finally, the matching conditions. First, note that we cannot naively match all the derivatives of l at the bubble wall even though doing so indeed gives the correct number of conditions. Rather, we match T, Q, and H at the bubble wall along with their first derivatives. Suppose $\gamma_i > 0$ for $i \in \{4,5,6\}$ and $\alpha_i > 0$ for $i \in \{1,2,3\}$. Let us use the convenient notation that $A_X(\alpha_i) \equiv A_X^{bi}$ and $A_X(\gamma_i) \equiv A_X^{si}$. The matching conditions result in the following conditions

$$\begin{aligned}
&\left(x_4\beta_4 \;\; x_5\beta_5 \;\; x_6\beta_6 \;\; x_1\beta_1 \;\; x_2\beta_2 \;\; x_3\beta_3 \;\; y_1 \;\; y_2 \;\; y_3\right)^T = \\
&= \begin{pmatrix}
1 & 0 & 0 & 0 & 0 & 0 & 0 & 0 & 0 \\
0 & 1 & 0 & 0 & 0 & 0 & 0 & 0 & 0 \\
0 & 0 & 1 & 0 & 0 & 0 & 0 & 0 & 0 \\
A_Q^{b4} & A_Q^{b5} & A_Q^{b6} & A_Q^{b1} & A_Q^{b2} & A_Q^{b3} & A_Q^{s1} & A_Q^{s2} & A_Q^{s3} \\
\alpha_4 A_Q^{b4} & \alpha_5 A_Q^{b5} & \alpha_6 A_Q^{b6} & \alpha_1 A_Q^{b1} & \alpha_2 A_Q^{b2} & \alpha_3 A_Q^{b3} & \gamma_1 A_Q^{s1} & \gamma_2 A_Q^{s2} & \gamma_3 A_Q^{s3} \\
A_T^{b4} & A_T^{b5} & A_T^{b6} & A_T^{b1} & A_T^{b2} & A_T^{b3} & A_T^{s1} & A_T^{s2} & A_T^{s3} \\
\alpha_4 A_T^{b4} & \alpha_5 A_T^{b5} & \alpha_6 A_T^{b6} & \alpha_1 A_T^{b1} & \alpha_2 A_T^{b2} & \alpha_3 A_T^{b3} & \gamma_1 A_T^{s1} & \gamma_2 A_T^{s2} & \gamma_3 A_T^{s3} \\
A_H^{b4} & A_H^{b5} & A_H^{b6} & A_H^{b1} & A_H^{b2} & A_H^{b3} & A_H^{s1} & A_H^{s2} & A_H^{s3} \\
\alpha_4 A_H^{b4} & \alpha_5 A_H^{b5} & \alpha_6 A_H^{b6} & \alpha_1 A_H^{b1} & \alpha_2 A_H^{b2} & \alpha_3 A_H^{b3} & \gamma_1 A_H^{s1} & \gamma_2 A_H^{s2} & \gamma_3 A_H^{s3}
\end{pmatrix}^{-1}
\begin{pmatrix} I_1 \\ I_2 \\ I_3 \\ 0 \\ 0 \\ 0 \\ 0 \\ 0 \\ 0 \end{pmatrix} \\
&\equiv \begin{pmatrix}
\mathbf{1}_{3\times 3} & 0 \\
A_X(\alpha) & A_X(\gamma) \\
(\alpha A)_X(\alpha) & (\alpha A)_X(\gamma)
\end{pmatrix}^{-1}
\begin{pmatrix} I_1 \\ I_2 \\ I_3 \\ 0 \end{pmatrix}.
\end{aligned} \tag{10.44}$$

We can now write the analytic form of the solutions for H, T, and Q:

$$
\begin{aligned}
H &= \sum_{i=1}^{6} A_H(\alpha_i) x_i e^{\alpha_i z} \left(\int^{z} e^{-\alpha_i y} \Delta(y) dy - \beta_i \right) \\
T &= \sum_{i=1}^{6} A_T(\alpha_i) x_i e^{\alpha_i z} \left(\int^{z} e^{-\alpha_i y} \Delta(y) dy - \beta_i \right) \\
Q &= \sum_{i=1}^{6} A_Q(\alpha_i) x_i e^{\alpha_i z} \left(\int^{z} e^{-\alpha_i y} \Delta(y) dy - \beta_i \right),
\end{aligned}
\tag{10.45}
$$

with known functions defined as

$$
\begin{aligned}
A_H &= - \sum_{n=0}^{4} \delta_{i+j-n} (a_{Q2}^{i} a_{T1}^{j} - a_{T2}^{i} a_{Q1}^{j}) \alpha^n \\
A_T &= - \sum_{n=0}^{4} \delta_{i+j-n} a_{Q1}^{i} a_{H2}^{j} \alpha^n \\
A_Q &= \sum_{n=0}^{4} \delta_{i+j-n} a_{T1}^{i} a_{H2}^{j} \alpha^n .
\end{aligned}
\tag{10.46}
$$

So we have derived a fairly simple solution to the transport equations for the MSSM case. However, this calculation is easily generalized to multiple transport equations. Also, it is significantly simpler to assume the form of the solutions, parametrized in terms of α, $A_X(\alpha)$, β, and x_i, and substitute back into the transport equations to obtain a set of conditions which specify the above parameters. Specifically, α and $A_X(\alpha)$ solve the homogeneous version of the transport equations and x_i is defined when substituting the full solution into the inhomogeneous transport equations. Furthermore, it is straightforward to deal with the case of multiple transport equations—simply replace x_i with x_{ij} and the sources $\Delta(z)$ become replaced with $\Delta_j(z)$. One can then proceed as described above, either deriving the solution or substituting the known form to derive the parameters.

10.5 Beyond ultrathin walls

When we approximate the mass relaxation terms as constant in the broken phase and zero in the other, we are linearizing our differential equations. Therefore, we can also write perturbations to the solution which take into account the space–time dependence of the mass terms. Consider the second transport equation (10.22), which contains mass terms proportional to a. Let us assume for simplicity that a is small so that the only mass terms are in equation (10.22). We will define the following error functions

$$\begin{aligned}
\Delta(z) &= \Theta(z)\Delta(z) + (1 - \Theta(z))\Delta(z) \equiv \Delta_0(z) + \epsilon(z) \\
a_{l3}^0(z) &\equiv a_{l3}^0(z)\Theta(z) + a_{l3}^0(z)\Theta(-z) \\
&= a_{l3}^0(z_{\max}) - [a_{l3}^0(z_{\max}) - a_{l3}^0(z)] + a_{l3}^0(z)\Theta(-z) \\
&= a_{l3}^0 + \delta a_{l3}^0(z) \\
l(z) &= l_0(z) + \delta_1 l(z) + \delta_2 l(z) + \cdots.
\end{aligned} \tag{10.47}$$

Note that $l_0(z)$ solves the linearized transport equations so we can make a dramatic simplification similar to the simplification we made when solving bubble wall profiles. Also, just like in the case of the bubble wall profiles, these perturbations are finite, which allows for a perturbative series. Inserting the perturbations into the transport equations gives

$$\begin{aligned}
a_{l3}^i \partial^i l_0 + a_{l3}^i \partial i(\delta_1 l + \delta_2 l + \cdots) + \delta a_{l3}^0(z)(l_0 + \delta_1 l + \delta_2 l + \cdots) \\
= \Delta(z) + \epsilon(z).
\end{aligned} \tag{10.48}$$

We can use the exact same method as before to find the corrections to m

$$\begin{aligned}
\delta_1 l &= \sum_{i=0}^{6} e^{\alpha_i z} x_i \left(\int^z e^{-\alpha_i y}[\epsilon - l_0 \delta a_{l3}^0(z)] - \delta_1 \beta_i \right) \\
\delta_2 l &= \sum_{i=0}^{6} e^{\alpha_i z} x_i \left(\int^z e^{-\alpha_i y}[-\delta_1 l \delta a_{l3}^0(z)] - \delta_2 \beta_i \right),
\end{aligned} \tag{10.49}$$

etc.

References

[1] Moore G D 2000 Sphaleron rate in the symmetric electroweak phase *Phys. Rev.* D **62** 8

[2] White G A 2016 General analytic methods for solving coupled transport equations: from cosmology to beyond *Phys. Rev.* D **93** 4

IOP Publishing

Electroweak Baryogenesis (Second Edition)
An introduction
Graham White

Chapter 11

The semi-classical force

So far, we have focused on the resonantly enhanced source that arises from interactions with the space–time varying vacuum. The calculation we have followed has its drawbacks. We only performed the first order in the vacuum expectation value (VEV) insertion diagrams without checking convergence (for a discussion of convergence see reference [1]). Second, we had to perform the calculation in the mass basis of the symmetric phase for the sake of tractability in order to derive a source that has support deep within the broken phase. Third, we assumed that we could remove the space–time varying vevs from the integral, (another assumption which was explored in reference [2]). The second approximation is particularly concerning because it is not obvious how to tell when this approximation breaks down. The severity of these drawbacks makes it worth considering a different approach.

The most well-known alternative approach to analyzing the CP violation is the semi-classical approach we will use in this chapter. It avoids a lot of the ad hoc simplifications present in the VEV insertion approach, but itself has some drawbacks. Most notably, the semi-classical force predicts a negligible source if there is CP violation in a scalar sector. Qualitatively, it is clear from a brute force numerical treatment of a scalar sector, that the CPV source is not negligible and peaks when there is a mass degeneracy as predicted in the VEV insertion approach [3]. Further, a resonantly enhanced source can be derived analytically in a simplified system even when the VEVs are resummed [4], though the relationship between this source and the one derived in the VEV insertion approach is to date a mystery. A summary of the pros and cons of the two approaches we give in table 11.1.

If there is CP violation in the mass sector of a theory, the fermion mass can be complex and spatially varying

$$M = m(z)e^{i\theta(z)}. \tag{11.1}$$

Table 11.1. Pros and cons of different approaches to calculating CPV sources.

VEV insertion approach		Semi-classical force	
Pros	Cons	Pros	Cons
• Focuses on the dominant source • Allows for exploring models with scalar sources	• Not all approximations are easy to check for self-consistency • Unless conditions are extremely unfavourable, it is difficult to make robust statements about the experimental status of a model	• Robust, derived from first principles involving only controlled approximations. • Only source when the mass matrix is 1×1.	• Might prematurely rule out a model • Cannot analyze a model with a scalar sector

The fermion will obey the Dirac equation with a complex, spatially varying mass

$$\left(i\gamma^\mu\partial_\mu - M\right)\psi = 0. \tag{11.2}$$

Let us boost to the rest frame of the bubble wall. Spin is conserved in both phases, allowing us to separate the chirality and spin degrees of freedom. A spin eigenstate is then

$$\Psi_s = e^{-i\omega t}\begin{pmatrix} L_s \\ R_s \end{pmatrix} \otimes \chi_s, \tag{11.3}$$

where L_S and R_S are the amplitudes for each chirality state and we choose our coordinates such that the spin wave function, χ_s, is an eigenstate of the spin-z operator. Inserting the spin eigenstate into the Dirac equation yields two coupled equations for the chirality amplitudes

$$(\omega - is\partial_z)L_s = MR_s, \tag{11.4}$$

$$(\omega + is\partial_z)R_s = M^* L_s. \tag{11.5}$$

We can eliminate the right-handed chirality amplitude from the second equation using the first. The result is a single second-order differential equation

$$(\omega - is\partial_z)\frac{1}{M}(\omega + is\partial_z)L_s = M^* L_s. \tag{11.6}$$

To solve this equation, we use the Wentzel–Kramers–Brillouin (WKB) ansatz

$$L_s \sim e^{i\int p_{cz}(z')dz'}, \tag{11.7}$$

where p_{cz} is the canonical momentum in direction of the expanding bubble wall. We then insert the WKB ansatz into our second-order differential equation and separate into real and imaginary components. It is straightforward to derive a dispersion relation that is correct to first order in gradients

$$\omega = \sqrt{(p_c - \alpha' + s_{CP}\theta'/2)^2 + m^2} - s_{CP}\frac{s\theta'}{2}, \tag{11.8}$$

where $s_{CP} = \pm 1$ for particles and antiparticles, respectively, and α is a gauge ambiguity that the Dirac equation is solved by any WKB ansatz related by the transformation $p_c \to p_c + \alpha$. The force acting on a particle is given by $F = p$ where p is related to the group velocity via $p = \omega v_g = \omega \frac{\partial \omega}{\partial p_c}$. With a little effort it is straightforward to derive the semi-classical force [5–7]

$$F = -\frac{(m^2)'}{\omega} + s_{CP}\frac{s(m\theta')'}{2\omega^2}. \tag{11.9}$$

This might at first seem to be as crude as the approximations appearing in the VEV insertion approach, as thus far we have not used a field theoretic approach. However, we can derive the same force in the CTP formalism treating propagators. We can begin with the collisionless Schwinger–Dyson equation in the CTP formalism

$$\left(i\gamma^\mu \partial_\mu \text{-} \hat{M}\right) iS^{ab} = a\delta_{ab} i\delta^4(x - y). \tag{11.10}$$

In the above, the spatially varying mass is given by a sum of the real and imaginary components of the mass

$$\hat{M} = m^R + i\gamma^5 m^I, \tag{11.11}$$

and the indices, a and b, denote the CTP indices $\in \pm$. Of course, without imaginary component, there is no CP violation. To solve the Schwinger–Dyson equation, we will perform a transformation such we need consider only a single spatial variable and a momentum variable (we can then integrate over momentum to derive moments). The transformation we will perform is known as the Wigner transform,

$$G(k, x) = \int \mathrm{d}^4 r e^{ik \cdot r} G\left(x + \frac{r}{2}, x - \frac{r}{2}\right), \tag{11.12}$$

After applying this transformation to the collisionless Dyson–Scwhinger equation for the '<' propagator, one arrives at the Kadanoff–Baym equation

$$\left[\gamma^\mu k_\mu + \frac{i}{2}\gamma^\mu \partial_\mu - m^R e^{-\frac{i}{2}\overleftarrow{\partial} \cdot \partial_k} - i\gamma^5 m^I e^{-\frac{i}{2}\overleftarrow{\partial} \cdot \partial_k}\right] S^< = 0. \tag{11.13}$$

From here, we could perform a full first principles calculation involving a helicity decomposition of the Wightman functions, deriving a constraint and kinetic equation at second order in gradients and then solving to derive the semi-classical force. However, this is a little gory and overkill to convince the reader that the semi-classical force can be derived from first principles in a field theoretic approach. We will follow a compromise here that sacrifices little in terms of rigor but gains a lot in simplicity. The interested reader can refer to references [4, 8] for a more rigorous approach.

We can manipulate the Kadanoff–Baym equation into an equation for the chiral current by multiplying by γ^5 and taking the trace. Using $j^{5z} = \langle \overline{\psi}\gamma^3\gamma^5\psi \rangle = iTr[\gamma^3\gamma^5 S^<]$ one finds that

$$-ik_z j^{5z} + \frac{1}{2}\partial_z j^{5z} - iTr\left[\left(m^R\gamma^5 e^{-\frac{i}{2}\partial\cdot\partial_k}\gamma^0 + im^I e^{-\frac{i}{2}\partial\cdot\partial_k}\gamma^0\right)i\gamma^0 S^<\right] = 0, \qquad (11.14)$$

It is straightforward to generalize this to a covariant form. However, since the bubble wall is moving in the z-direction, we can expect that the only chiral current that is sourced through CP violating interactions with the bubble wall will be j^{5z}. We will, therefore, simplify our analysis by ignoring the other terms. Note that we have inserted factors of γ^0 in order to put the equation into a format where it is easy to take the Hermitian conjugate, as $Tr[i\gamma^0 S^<]^\dagger = n^\dagger = Tr[i\gamma^0 S^<]$. This allows us to derive the equation

$$\partial_z j^{5z} + m^R\left(e^{-\frac{i}{2}\partial\cdot\partial_k} + e^{\frac{i}{2}\partial\cdot\partial_k}\right)Tr[\gamma^5 S^<] + im^I\left(e^{-\frac{i}{2}\partial\cdot\partial_k} + e^{\frac{i}{2}\partial\cdot\partial_k}\right)Tr[S^<] = 0. \qquad (11.15)$$

We can similarly obtain equations for $Tr[\gamma^5 S^<]$ and $Tr[S^<]$ by multiplying the Kadanoff–Baym equation by γ^3 and $\gamma^3\gamma^5$ from the left, respectively, then adding the Hermitian and anti-Hermitian conjugates, respectively. The results are

$$Tr[S^<] = \frac{1}{2k_z}m^R\left(e^{-\frac{i}{2}\partial\cdot\partial_k} + e^{\frac{i}{2}\partial\cdot\partial_k}\right)Tr[\gamma^3 S^<] + \frac{i}{2k_z}m^I\left(e^{-\frac{i}{2}\partial\cdot\partial_k} - e^{\frac{i}{2}\partial\cdot\partial_k}\right)j^{5z}, \qquad (11.16)$$

$$Tr[\gamma^5 S^<] = -\frac{i}{2k_z}m^I\left(e^{-\frac{i}{2}\partial\cdot\partial_k} + e^{\frac{i}{2}\partial\cdot\partial_k}\right)Tr[\gamma^3 S^<] + \frac{1}{2k_z}m^R\left(e^{-\frac{i}{2}\partial\cdot\partial_k} - e^{\frac{i}{2}\partial\cdot\partial_k}\right)j^{5z}. \qquad (11.17)$$

Substituting these equations for the traces of the Wightman functions back into our equation for our chiral current in the z-direction,

$$\begin{aligned}
&\partial_z j^{5z} + m^R\left(e^{-\frac{i}{2}\partial\cdot\partial_k} + e^{\frac{i}{2}\partial\cdot\partial_k}\right)\left(-\frac{i}{2k_z}m^I\left(e^{-\frac{i}{2}\partial\cdot\partial_k} + e^{\frac{i}{2}\partial\cdot\partial_k}\right)Tr[\gamma^3 S^<]\right. \\
&\left. + \frac{1}{2k_z}m^R\left(e^{-\frac{i}{2}\partial\cdot\partial_k} - e^{\frac{i}{2}\partial\cdot\partial_k}\right)j^{5z}\right) + im^I\left(e^{-\frac{i}{2}\partial\cdot\partial_k} + e^{\frac{i}{2}\partial\cdot\partial_k}\right) \\
&\left(\frac{1}{2k_z}m^R\left(e^{-\frac{i}{2}\partial\cdot\partial_k} + e^{\frac{i}{2}\partial\cdot\partial_k}\right)Tr[\gamma^3 S^<] + \frac{i}{2k_z}m^I\left(e^{-\frac{i}{2}\partial\cdot\partial_k} - e^{\frac{i}{2}\partial\cdot\partial_k}\right)j^{5z}\right) = 0.
\end{aligned} \qquad (11.18)$$

So far, our fluid equation is exact. To derive the semi-classical force, let us assume a power counting which attributes a power of ϵ for each power of gradients and the correction of $Tr[\gamma^3 S^<]$ from its equilibrium value to also be a non-zero power of ϵ. Expanding to ϵ^2, and focusing on terms containing $Tr[\gamma^3 S^<]$ we have

$$\frac{i}{2}\left([\partial^2 m^R]m^I - [\partial^2 m^I]m^R\right)\left(\frac{1}{k_z}\partial_k\, Tr[\gamma^3 S^<]\right). \qquad (11.19)$$

It is straightforward to show that $\frac{i}{2}([\partial^2 m^R]m^I - [\partial^2 m^I]m^R) = (m\theta')'$, this has the same spatial dependence as the source appearing in the second term in equation (11.9). The first term in equation (11.9) is a CP conserving classical force which can be found in the field theoretic approach by isolating terms proportional to j^{5z}

$$\frac{1}{2}(m^2)'\frac{1}{k_z}\partial_k j^{5z}. \tag{11.20}$$

References

[1] Postma M and Van De Vis J 2019 Source terms for electroweak baryogenesis in the vev-insertion approximation beyond leading order *J. High Energy Phys.* **02** 090

[2] Postma M 2021 A different perspective on the vev insertion approximation for electroweak baryogenesis *J. High Energy Phys.* **09** 055

[3] Cirigliano V, Lee C and Tulin T 2011 Resonant flavor oscillations in electroweak baryogenesis *Phys. Rev.* D **84** 056006

[4] Garbrecht B 2018 Why is there more matter than antimatter? Calculational methods for leptogenesis and electroweak baryogenesis *Prog. Part. Nucl. Phys.* **110** 103727

[5] Joyce M, Pokopec T and Turok N 1995 Electroweak baryogenesis from a classical force *Phys. Rev. Lett.* **75** 3375

[6] Joyce M, Prokopec T and Turok N 1996 Nonlocal electroweak baryogenesis. Part 2: the classical regime *Phys. Rev.* D **53** 2958

[7] Cline J and Kainulainen K 2000 A new source for electroweak baryogenesis in the MSSM *Phys. Rev. Lett.* **85** 5519

[8] Kainulainen K, Prokopec T, Schmidt M and Weinstock S 2001 First principle derivation of semiclassical force for electroweak baryogenesis *J. High Energy Phys.* **06** 031

IOP Publishing

Electroweak Baryogenesis (Second Edition)
An introduction
Graham White

Chapter 12

The semi-classical force in fluid equations

Recall that for VEV insertion approach we derived a source for the divergence of a current, which was in turn converted into a Boltzmann equation for number densities via Fick's law. This approach, while containing some crude steps, has the advantage of allowing a straightforward path from the calculation of both the sources and the collision terms and deriving a set of Boltzmann equations. The derivation of the Boltzmann equations in the semi-classical force picture requires its own chapter. It is typical [1, 2] to consider an Ansatz for the distribution function that includes perturbations from the equilibrium distribution,

$$f = \frac{1}{e^{\beta(\gamma(E+v_w p_z)-\mu)} \pm 1} + \delta f, \tag{12.1}$$

and separate both the chemical potential and the perturbation of the distribution function into CP even and CP odd terms

$$\mu = \mu_e + s_c \mu_o, \tag{12.2}$$

$$\delta f = \delta f_e + \delta f_o. \tag{12.3}$$

Here the subscripts '*e,o*' refer to the even and odd contributions, respectively. The energy can be also written as its distribution plus the leading order term in gradients

$$E = E_0 + \Delta E, \tag{12.4}$$

where $\Delta E = \frac{\theta' m^2}{2E_0 E_{0z}}$, $E_0 = \sqrt{m^2 + p^2}$ and $E_{0z} = \sqrt{m^2 + p_z^2}$. Expanding the distribution function to leading order in all small parameters,

$$f = f_0 + \left[-\mu_e f_0' + \delta f_e\right] + \left((-\mu_o + s\gamma \Delta E) f_0' - s_h \gamma \Delta E f_0'' \mu_e + \delta f_o\right), \tag{12.5}$$

where we have put the CP even (odd) in square (circular) brackets and f_0 is the equilibrium distribution function. We can then write the usual Boltzmann equation for the distribution function

$$(v_g\partial + F\partial_k)f = C[f]. \tag{12.6}$$

This of course is just a classical Boltzmann equation where the right-hand side is the collision term and the left hand side can be derived trivially using the chain rule, $\frac{df}{dt} = \vec{v}\cdot\nabla f + \frac{d\vec{k}}{dt}\cdot\nabla_k f$. We can then derive a set of fluid equations simply by inserting the distribution function written in equation (12.5) into the Boltzmann equation (12.6) and inserting the force term from the previous chapter. This leads to a formidable equation, though in principle one that can be solved. It is more typical to simplify the task slightly by taking moments of equation (12.6) and solving the new set of differential equations which only contain a spatial dependence. Defining

$$L = (v_g\partial + F\partial_k)f, \tag{12.7}$$

We wish to calculate and $\langle L\rangle$, $\langle\frac{p_z}{E_0}L\rangle$. Defining,

$$u_l = \left\langle \left(\frac{p_z}{E_0}\right)^l \delta f \right\rangle,\quad D_l = \left\langle \left(\frac{p_z}{E_0}\right)^l f_0' \right\rangle,\quad Q_l = \left\langle \frac{p_z^{l-1}}{2E_0^l} f_0'' \right\rangle,$$

the moments of the Louisville operator have the form

$$\langle L\rangle = -D_1\mu' + u_1' + v_w\gamma\, m^{2'} Q_1\mu + \text{CPV terms}, \tag{12.8}$$

$$\left\langle \frac{p_z}{E_0}L \right\rangle = -D_2\mu' + u_2' + v_w\gamma\, m^{2'} Q_2\mu + m^{2'}\left\langle \frac{1}{2E_z^2}\delta f \right\rangle + \text{ CPV terms}. \tag{12.9}$$

We have for now neglected the CP violating terms. The set of equations has two problematic terms, the first is u_2 which is higher order in the expansion than what we are working to, and the last term where the connection to velocity perturbations isn't clear. Following reference [1], we will assume a factorization scheme

$$u_2 = \left\langle \left(\frac{p_z}{E_0}\right)^2 \delta f \right\rangle \sim \left[\frac{p_z}{E_0}\right]\left\langle \frac{p_z}{E} \right\rangle \delta f = \left[\frac{p_z}{E_0}\right] u = Ru,\quad \left\langle \frac{1}{2E_z^2}\delta f \right\rangle = \left[\frac{1}{2p_z E_0}\right] u = \bar{R}u, \tag{12.10}$$

where the square bracket average is $[\,\cdot\,] = \int d^3x(\,\cdot\,)$. The set of differential equations (12.8) and (12.9) can be written in a matrix form

$$Aw' + m^{2'}Bw = \bar{S} + \delta C, \tag{12.11}$$

with

$$w = \begin{pmatrix}\mu\\ u\end{pmatrix},\quad A = \begin{pmatrix}-D_1 & 1\\ -D_2 & R\end{pmatrix},\quad B = \begin{pmatrix}v_w\gamma Q_1 & 0\\ v_w\gamma Q_2 & \bar{R}\end{pmatrix},\quad \bar{S} = \begin{pmatrix}\langle S\rangle\\ \left\langle \frac{p_z}{E_0}S \right\rangle\end{pmatrix}, \tag{12.12}$$

where we will neglect the collision term for now. The CP even source arise from the first term in the force we derived in the previous chapter and the CP odd source is the semi-classical force,

$$S_l^e = v_w\gamma\, m^{2'} Q_l^e, \tag{12.13}$$

$$S_{hl}^o = -v_w\gamma\, h(m^2\theta')' Q_l^o. \tag{12.14}$$

In the above we have used the shorthand $Q_l^e = \left\langle \frac{p_z^{l-1}}{2E^l} f_0' \right\rangle$ and $Q_l^o = \left\langle \frac{s_p p_z^{l-1}}{2E^l E_z} f_0' \right\rangle$ and we have neglected a term that is often numerically subdominant.

It is worth seeing an explicit example. Let us consider an explicit example which is still a viable model of electroweak baryogenesis. Consider a source of CP violation in the top quark sector

$$m_t(z)(\bar{t}_L e^{i\theta(z)} t_R + \bar{t}_R e^{-i\theta(z)} t_L) \tag{12.15}$$

where the CP violating phase can arise from interference between the Yukawa interaction a higher dimensional operator. The set of differential equations governing the CP odd sector has the form

$$A_t w_{t_L}' + m_t^{2'} B_t w_{t_L} - \delta C_{t_L} = S_t, \tag{12.16}$$

$$A_b w_{b_L}' + m_b^{2'} B_b w_{b_L} - \delta C_{b_L} = 0, \tag{12.17}$$

$$A_t w_{t_R}' + m_t^{2'} B_t w_{t_R} - \delta C_{t_R} = -S_t, \tag{12.18}$$

$$A_h w_h' + m_h^{2'} B_h w_h - \delta C_h = 0. \tag{12.19}$$

The first collision terms have the same form as in the previous chapters,

$$\delta\overline{C}_1^{t_L} = \Gamma_y(\mu_{t_L} - \mu_{t_R} + \mu_h) + \Gamma_m(\mu_{t_L} - \mu_{t_R}) + \Gamma_W(\mu_{t_L} - \mu_{b_L}) + \Gamma_{SS}\mu_5,$$

$$\delta\overline{C}_1^{b} = \Gamma_y(\mu_{b_L} - \mu_{b_R} + \mu_h) - \Gamma_W(\mu_{t_L} - \mu_{b_L}) + \Gamma_{SS}\mu_5,$$

$$\delta\overline{C}_1^{t_L} = \Gamma_y(\mu_{t_L} + \mu_{b_L} - 2\mu_{t_R} + \mu_h) + \Gamma_m(\mu_{t_L} - \mu_{t_R}) - \Gamma_{SS}\mu_5,$$

$$\delta\overline{C}_1^{t_L} = \Gamma_y(\mu_{t_L} + \mu_{b_L} - 2\mu_{t_R} + \mu_h) + \Gamma_h\mu_h. \tag{12.20}$$

Finally, the second set of collision terms are related to the first via the relation $\delta C_2^i = -\Gamma_{tot}^i u_i + v_w\langle f_0\rangle\delta C_1^i$.

Early work calculating the baryon asymmetry in the Wentzel–Kramers–Brillouin approach expanded to leading order in the bubble wall velocity. In the leading order calculations of reference [1], the baryon asymmetry vanishes near the speed of sound. However, recent work [2, 3] has considered going beyond the leading order expansion and has found it is possible to have viable baryon production with a large bubble wall velocity. This is fortuitous, as a large wall velocity is associated with a stronger phase transition and only the strongest phase transitions can be conceivably detected by next generation gravitational wave detectors. Unfortunately, it seems that at least for the simplest model it is difficult for a set of parameters consistent with electroweak baryogenesis to yield a detectable signal.

References

[1] Fromme L and Huber S 2007 Top transport in electroweak baryogenesis *J. High Energy Phys.* **03** 049

[2] Cline J and Kainulainen K 2020 Electroweak baryogenesis at high bubble wall velocities *Phys. Rev.* D **101** 6

[3] Cline J, Friedlander A, He D, Kainulainen K and Laurent B 2021 Baryogenesis and gravity waves from a UV-completed electroweak phase transition *Phys. Rev.* D **103** 12

Chapter 13

The baryon asymmetry

Finally, let us perform the calculation, then put everything together and use the solutions to our transport equations to calculate the baryon asymmetry of the Universe. Weak sphalerons interact with the total left-handed number density to produce a baryon asymmetry. From the solutions to the transport equations, one can then define the left-handed number density $n_L(z) = Q_{1L} + Q_{2L} + Q_{3L} = 5Q + 4T$. The baryon number density, ρ_B, satisfies the equation for a certain relaxation constant R [1, 2]

$$D_Q\rho_B''(z) - v_W\rho_B'(z) - \Theta(-z)\mathcal{R}\rho_B = \Theta(-z)\frac{n_F}{2}\Gamma_{ws}n_L(z), \tag{13.1}$$

where n_F is the number of fermion families. The relaxation parameter can be derived by asserting that

$$A\frac{n_B}{T^2} = \mu_{\rm CS} = 9\mu_q + \mu_{l_i}, \tag{13.2}$$

where $\mu_{\rm CS}$ is the chemical potential for the Chern–Simons number for the $SU(2)_L$ gauge field. The second equation in the above comes from the fact that a weak sphaleron creates nine quarks and three leptons. We can assume quark mixing interactions are fast compared to the weak sphaleron rate in the symmetric phase. We can then write for the Standard Model

$$n_B = \frac{1}{3}n_q \to \mu_q = \frac{1}{2}\frac{\mu_q}{T^2} \tag{13.3}$$

and the lepton number can be related to the baryon number via $\sum_i\mu_{l_i} = 2n_B/T^2$ under the assumption that the right-handed leptons do not have time to equilibrate which gives $A = 15/2$. For the MSSM a simple calculation reveals

doi:10.1088/978-0-7503-3571-3ch13

$$\mathcal{R} = \Gamma_{ws}\left[\frac{9}{4}\left(1 + \frac{n_{sq}}{6}\right) + \frac{3}{2}\right] \tag{13.4}$$

where n_{sq} is the number of squarks thermally available at the nucleation temperature and $\Gamma_{ws} \approx 120\alpha_W^5 T$ [3]. The baryon asymmetry of the Universe, Y_B is then given by

$$Y_B = -\frac{n_F \Gamma_{ws}}{2\kappa_+ D_Q S} \int_{-\infty}^{0} e^{-\kappa_- x} n_L(x) dx, \tag{13.5}$$

where

$$\kappa_\pm = \frac{v_W \pm \sqrt{v_W^2 + 4D_Q \mathcal{R}}}{2D_Q} \tag{13.6}$$

and the entropy is given by

$$S = \frac{2\pi^2}{45} g_* T^3. \tag{13.7}$$

References

[1] Moreno J M, Quiros M and Seco M 1998 Bubbles in the supersymmetric standard model *Nucl. Phys.* B **526** 489–500

[2] Cline J M, Joyce M and Kainulainen K 2000 Supersymmetric electroweak baryogenesis *J. High Energy Phys.* **JHEP07** 018

[3] Bödeker D, Moore G D and Rummukainen K 2000 Chern–Simons number diffusion and hard thermal loops on the lattice *Phys. Rev.* D **61** 5

Chapter 14

A brief phenomenological summary

In this chapter we give a brief review of the current status of models. At present, the recent ACME results [1] give a strong constraint on any model predicting a permanent electron dipole moment larger than the Standard Model. The current constraint is,

$$d_e < 1.1 \times 10^{-29} \text{e cm} \tag{14.1}$$

Even still, the simplest models of electroweak baryogenesis are still allowed, with the caveat that robust statements are subject to the large uncertainties in the CPV source. Of the simplest models, the most pessimistic case is that of the minimal supersymmetric Standard Model (MSSM) where electroweak baryogenesis is ruled out apart from tiny corners of the parameter space [2, 3]. It is plausible that updates to this picture will rule it out completely. The key issue is that it is difficult to hide stops that are light enough to catalyze a strong transition. However, the next-to-minimal MSSM (NMSSM) is still quite viable, both in producing a strong first-order transition through use of the additional singlet [4, 5] and sufficient CP violation is possible through singlino interactions [6–8].

For non-SUSY models, minimal models include the Standard Model Effective Theory [9] the two-Higgs doublet model [10] a singlet extension [11] and the extension by a vector like fermion [12]. All are still viable models, although next generation electrid dipole moments (EDMs) will test the Standard Model Effective Theory and the two-Higgs doublet model currently requires a minor fine tuning in order to achieve a cancellation in the EDM prediction. Of course, non-minimal models are likely to be less constrained. One can also remove the constraint of electric dipole moments considerably through a slight modification of the overall paradigm. In terms of the Sakharov conditions,

1. One can consider the possibility that B number violation was more efficient in the early Universe—e.g. if the SU(2) gauge coupling constant was larger [13];

doi:10.1088/978-0-7503-3571-3ch14

2. One can modify how one departs from equilibrium—e.g. if the electroweak phase transition occurred through a multi-step phase transition [14–16];
3. One can consider CP violation to be different in the early Universe compared today—e.g. if the Standard model CP violation was large in the early Universe due to time dependent Yukawa couplings [17].

Even if EDMs do not constrain a particular model, new states are required to catalyze a strong first-order electroweak phase transition. In fact, even if the electroweak phase transition is at a higher scale, one still likely requires new states at the electroweak scale cause a departure from the standard picture of a crossover at O(100) GeV[1]. Two test scenarios that are particularly used are

1. The nightmare scenario [19] where there is again a singlet extension to the Standard Model. But there is first a discrete symmetry to forbid mixing, as well as a lower bound on the mass such that the Higgs cannot decay into it, $m_s > m_H/2$.
2. A singlet extension to the Standard Model with no discrete symmetries [20]. This is often the canonical test scenario used in the literature.

These two scenarios are typically used to test the reach of future colliders. It is widely believed, that perhaps one of the only fundamental questions that can be tested at future colliders is what is the nature of electroweak symmetry breaking? In the absence of a discrete symmetry, the singlet typically mixes with the Standard Model like Higgs and decays and the parameter space that allows a strong first-order electroweak phase transition has a singlet mass of below 800 GeV. Thus, the singlet inherits the decay modes of the Higgs boson. A 100 TeV collider working at 30 ab^{-1} in luminosity can verify or completely exclude the parameter space that allows a strong first-order transition through di-Higgs production [20] and vector boson production [21].

For the nightmare scenario, most of the parameter space is out of reach for direct detection at a similar strength collider [22] (see also reference [23] for discussion of the nightmare scenario at a muon collider). The most promising method for testing this scenario is through measuring the triple Higgs coupling—which needs to vary by 30% from the Standard Model expectation to catalyze a strong first-order electroweak phase transition. The triple Higgs coupling can be measured direction through di-Higgs production. A 100 TeV collider with 30 ab^{-1} in luminosity is expected to be sensitive to deviations to the Standard Model triple Higgs coupling that are larger than 10% at 2σ. Thus, the nightmare scenario can at least be ruled out by future colliders.

Finally, a strong first-order electroweak phase transition predicts a stochastic gravitational wave background that should be visible today, given a sensitive enough gravitational wave detector. As discussed briefly in chapter 12, the stronger the phase transition, the more pressure accelerating the bubble wall. If the bubble

[1] An exception to this is if there is a large lepton asymmetry [18].

wall velocity is very fast, the efficiency of producing a baryon asymmetry is suppressed. However, the stronger the transition, the larger the stochastic gravitational wave background. At least for the simplest models, there appears to be a tension between the parameter space that allows electroweak baryogenesis and the parameter space that allows an observable gravitational wave background. Even still, gravitational wave cosmology in the near future has promise as an exclusion channel—any observation is likely to rule out electroweak background, or at least severely limit the possible compatible scenarios.

All these statements about the testability of electroweak baryogenesis is what makes the paradigm perhaps the most attractive scenario for explaining the asymmetry between matter and anti-matter. However, all these statements are subject to severe theoretical uncertainties. Most of the parameter space that predicts a first-order electroweak phase transition in a singlet extension to the Standard Model, does so for only certain values of the RG scale using conventional methods [21–24]. The fact that the gravitational wave amplitude can vary by many orders of magnitude due to the same infrared uncertainties also means that the degree of tension between an observable signal and electroweak baryogenesis is unknown. Finally, the amount of asymmetry for a given amount of CP violation varies by many orders of magnitude [24], which makes robust statements of the reach of EDMs difficult to discern. It seems that the theoretical progress over the next two decades will need to match the experimental effort.

References

[1] ACME Collaboration 2018 Improved limit on the electric dipole moment of the electron *Nature* **562** 7727

[2] Curtin D, Jaiswal P and Meade P 2012 Excluding electroweak baryogenesis in the MSSM *J. High Energy Phys.* **8** 1–27

[3] Liebler S, Profumo S and Stefaniak T 2016 Light stop mass limits from higgs rate measurements in the MSSM: is MSSM electroweak baryogenesis still alive after all? *J. High Energy Phys.* **04** 143

[4] Athron P, Balazs C, Fowlie A, Pozzo G and White G 2019 Strong first-order phase transitions in the NMSSM—a comprehensive survey *J. High Energy Phys.* **11** 151

[5] Kozaczuk J, Profumo S, Stephenson H and Wainwright C 2015 Cosmological phase transitions and their properties in the NMSSM *J. High Energy Phys.* **01** 144

[6] Cheung K, Hou T, Lee J and Senaha E 2012 Singlino-driven electroweak baryogenesis in the next-to-MSSM *J. High Energy Phys.* **710** 188

[7] Bian L, Guo H and Shu J 2018 Gravitational Waves, baryon asymmetry of the Universe and electric dipole moment in the CP violating NMSSM *Chin. Phys.* C **42** 9

[8] Sujeet A *et al* 2017. Electroweak baryogenesis in the $\mathbb{Z}_3$-invariant NMSSM *J. High Energy Phys.* **11** 1–26

[9] Vries J de, Postma M and van de Vis J 2019 The role of leptons in electroweak baryogenesis *J. High Energy Phys.* **4** 1–41

[10] Fuyuto K, Hou W-S and Senaha E 2020 Cancellation mechanism for the electron electric dipole moment connected with the baryon asymmetry of the Universe *Phys. Rev.* D **101** 011901

[11] Laurent B, Cline J M, Friedlander A, He D-M, Kainulainen K and Tucker-Smith D 2021 Baryogenesis and gravity waves from a UV-completed electroweak phase transition *Phys. Rev.* D **103** 123529
[12] Egana-Ugrinovic D 2017 The minimal fermionic model of electroweak baryogenesis. *J. High Energy Phys.* **12** 1–37
[13] Ellis S, Ipek S and White G 2019 Electroweak baryogenesis from temperature varying couplings *J. High Energy Phys.* **08** 002
[14] Inoue S, Ovanesyan G and Ramsey-Musolf M 2015 Two-step electroweak baryogenesis *Phys. Rev.* D **93** 015013
[15] Ramsey-Musolf M, Winslow P and White G 2018 Color breaking baryogenesis *Phys. Rev.* D **97** 12
[16] Baldes I and Servant G 2018 High scale electroweak phase transition: baryogenesis and symmetry non-restoration *J. High Energy Phys.* **10** 053
[17] Bruggisser S, Konstandin T and Servant G 2017 CP-violation for electroweak baryogenesis from dynamical CKM matrix *JCAP* **11** 034
[18] Schwarz D and Stuke M 2009 Lepton asymmetry and the cosmic QCD transition *JCAP* **11** 025
[19] Curtin D, Meade P and Yu C 2014 Testing electroweak baryogenesis with future colliders *J. High Energy Phys.* **11** 127
[20] Ramsey-Musolf M 2020 The electroweak phase transition: a collider target *J. High Energy Phys.* **09** 179
[21] Kotwal A, R Musolf M, No J and Winslow P 2016 Singlet-catalyzed electroweak phase transitions in the 100 TeV frontier *Phys. Rev.* D **94** 3
[22] Papaefstathiou A and White G 2021 The electro-weak phase transition at colliders: confronting theoretical uncertainties and complementary channels *J. High Energy Phys.* **05** 099
[23] Al Ali H *et al* 2021 The Muon Smasher's guide arXiv:2103.14043
[24] Cline J M and Laurent B 2021 Electroweak baryogenesis from light fermion sources: a critical study *Phys. Rev.* D **104** 083507

IOP Publishing

Electroweak Baryogenesis (Second Edition)
An introduction
Graham White

Chapter 15

Other mechanisms for producing the baryon asymmetry

Here we have reviewed the vanilla electroweak baryogenesis framework with an emphasis on the theoretical background. Many other mechanisms have been proposed for producing the baryon asymmetry of the Universe (BAU) (for a recent example in the electroweak framework or thereabouts see [1–12], and [13–18] for recent examples outside the electroweak baryogenesis framework). In this section we briefly review three other prominent paradigms for producing the BAU:

- *The Affleck–Dine model and leptogenesis.* These models belong to a different paradigm to making use of phase transitions. In the case of leptogenesis the departure from equilibrium occurs when the temperature drops below the mass of a heavy particle, which decays into lighter particles through lepton violating interactions. In the case of the Affleck–Dine mechanism, scalar particles carrying baryon number interact with the inflation during the reheating period after inflation. Once again the heavy particle decays into leptons and baryons.
- *The Peskin–Anderson mechanism.* Another paradigm to the above, this model utilizes cosmic inflation itself as the departure from equilibrium. Since inflation generally washes out any baryon asymmetry, making use of it—and thereby turning an enemy into a friend—was something of an interesting paradigm shift. It makes use of the gravitational anomaly to violate baryon number conservation.

Unfortunately, not all of these models will be testable in the foreseeable future, but they are all theoretically elegant and are a demonstration of the variety of answers there might be to the question of how the asymmetry between particles and anti-particles arose.

doi:10.1088/978-0-7503-3571-3ch15

15.1 Leptogenesis

The only particle experiment to convincingly show a departure from the Standard Model is the detection of nuetrino masses [19]. Their mass appears to be unnaturally small compared to other Standard Model particles. A mechanism for producing their mass as well as naturally ensuring that they are small is the see-saw mechanism. This mechanism works by adding a Majorana mass to the Standard Model, which is allowed since the right-handed neutrino is a gauge singlet [20]:

$$\mathcal{L}_\nu \ni y_e^{ij}\bar{e}_i L_j H^\dagger + y_\nu^{ij}\nu_i L_j H - \frac{M_i}{2}\nu_i\nu_i. \tag{15.1}$$

The Lagrangian mass parameter M_i is a free parameter and it is expected to be GUT scale. The two mass eigenstates involve a heavy one around the scale of M and one suppressed by ν^2/M. With order 1 couplings and a GUT scale of 10^{15} GeV one obtains a light neutrino mass of about 0.01 eV.

Apart from providing a natural mechanism for small neutrino masses, these mass terms in the Lagrangian violate the lepton number explicitly. When the temperature of the Universe is larger than M there is an abundance of right-handed neutrinos. The departure from equilibrium occurs when the temperature of the Universe drops below M and these particles decay into Higgs and left-handed leptons. The Yukawa couplings between the Higgs and the neutrino are complex and are the source of CP violation as the tree-level decay interferes with the one-loop decay, as shown in figure 15.1. The difference between the decay to leptons and anti-leptons is denoted by epsilon [21–23]

$$\epsilon = \frac{\Gamma(\nu \to H + l) - \Gamma(\nu \to H^\dagger + l^\dagger)}{\Gamma(\nu \to H + l) + \Gamma(\nu \to H^\dagger + l^\dagger)}. \tag{15.2}$$

For a similar mass hierarchy to the quark hierarchy one finds that $\epsilon \in [10^{-5}, 10^{-6}]$.

Some of the lepton asymmetry is then passed to the baryon sector through weak sphaleron interactions. To see how such a transfer takes place we make use of the fact that weak sphalerons violate $B + L$ but conserve $B - L$. If one assumes all Standard Model Yukawa and gauge interactions are fast enough to set the combination of chemical potentials involved in each subsequent reaction in equilibrium, one has the following expression for the baryon and lepton numbers in terms of Standard Model particles:

$$\begin{aligned} \mu_B &= \sum_i 2\mu_{Q_i} + \mu_{u_i} + \mu_{d_i} \\ \mu_L &= \sum_i 2\mu_{L_i} + \mu_{e_i}. \end{aligned} \tag{15.3}$$

If we apply the conservation of hypercharge and set the flavor changing interactions to equilibrium, one finds that

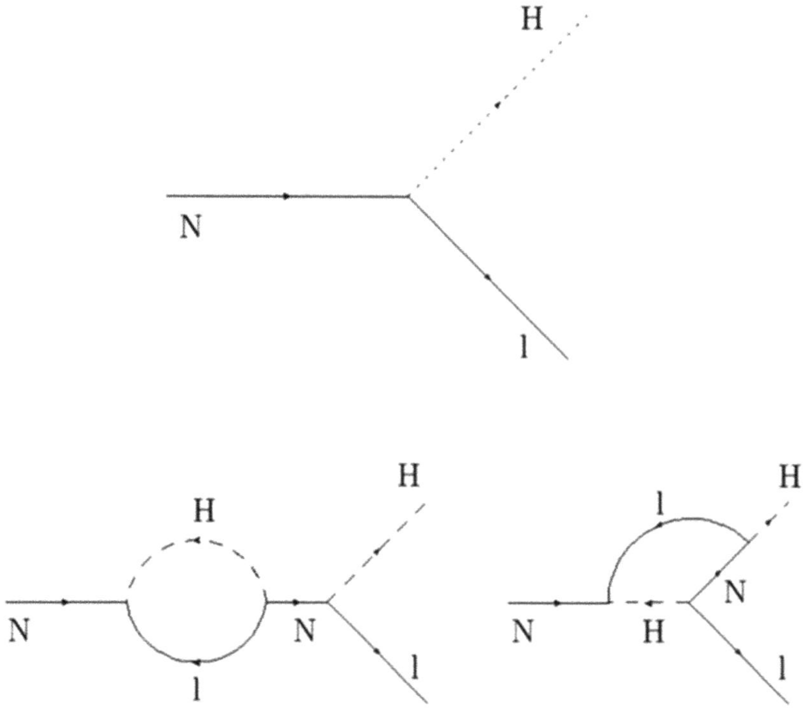

Figure 15.1. Tree and one-loop decay of the heavy neutrino. The CP violating phases governing the process ensures that the interference between tree and one-loop effects will lead to an asymmetry between leptons and anti-leptons.

$$
\begin{aligned}
B &= \frac{28}{79}(B - L) \\
L &= \frac{51}{79}(B - L).
\end{aligned}
\tag{15.4}
$$

We can therefore calculate the baryon asymmetry from the lepton asymmetry

$$
Y_B = \frac{n_B}{S} \approx -\frac{28}{79}\frac{n_L}{S} \approx 10^{-3}\eta\epsilon, \tag{15.5}
$$

where η is an efficiency factor that comes from solving the set of Boltzmann equations. For $\eta\epsilon \approx 10^{-7}$ one obtains the correct baryon asymmetry.

Leptogenesis is very simple and elegant but it is not without problems. The large Majorana mass causes a very severe fine-tuning problem (this of course is the usual hierarchy problem). If one solves such a hierarchy problem with SUSY then one runs into the problem that without some fine-tuning there is an overabundance of gravitinos. The biggest problem with leptogenesis, however, at least from the perspective of phenomenologists, is that the particles involved are so heavy (or alternatively so weakly coupled) that the scenario is untestable in the near future. One can indirectly corroborate it by tests of neutrinoless double beta decay which could, in principle, establish the Majorana nature of neutrinos. Assuming such an endeavor is successful, constraints on viable leptogenesis models can come through a measurement of the lepton asymmetry of the Universe (and comparing it to the BAU), as well as cosmic constraints on the neutrino mass sum as well as the masses of light neutrinos. This paradigm is quite high scale, but if $B - L$ is associated with a gauge symmetry that is spontaneously broken, it is more likely than not that the symmetry breaking produces cosmic strings which should leave a gravitational wave signature [24].

15.2 Affleck–Dine

The Affleck–Dine mechanism is similar to the leptogenesis approach in that it involves decays of heavy particles [25, 26]. In this case though the mechanism is a little more subtle, hence giving it separate treatment. During inflation, scalar fields with a shallow potential can acquire vacuum expectation values. SUSY, for example, has many such directions which can be scalar fields, ϕ, with non-zero lepton number [29]. It is possible that such a field could have a large vacuum expectation value (VEV) during inflation and both C and CP are spontaneously violated when this happens. The large Hubble parameter would act as a very strong friction term that prevents the condensate from decaying during inflation. After inflation the field rotates in field space and the condensate decays into fermions with non-zero lepton number. Consider an effective potential of the form

$$V(\phi) = (m_\phi^2 + c_H H^2)\phi^2 + O(\phi^4) + \cdots. \tag{15.6}$$

We have just written the mass term due to it being the term of most interest for this discussion. Consider the case where $c_H < 0$. The Hubble parameter is very large during inflation and the field will have a large VEV. Once inflation stops, the Hubble parameter decreases until the minimum of the potential is at the origin and the condensate will decay. The equation of motion for ϕ is

$$\ddot{\phi} + 3H\dot{\phi} + \frac{\partial V}{\partial \phi^*} = 0. \tag{15.7}$$

Since ϕ has lepton number one can write the lepton number density as

$$n_L = i\beta(\dot{\phi}\phi^* - \phi\dot{\phi}^*). \tag{15.8}$$

Using the above two equations we have

$$\dot{n}_L + 3Hn_L + 2\beta \,\mathrm{Im}\left[\frac{\partial V}{\partial \phi^*}\phi^*\right]. \tag{15.9}$$

The field during reheating spirals into the origin, which means a non-zero lepton number throughout. This can decay into a permanent lepton number due to couplings between the scalar field and Standard Model leptons. The lepton asymmetry can then be transferred to a baryon asymmetry through electroweak sphalerons. Like leptogenesis, the physics involved is often at very high scales.

The most promising avenue for testing it is through gravitational waves that are enhanced by a sudden change in equation of state that is generically predicted by this paradigm [30].

15.3 Using inflation

How to complete the Standard Model to include gravity in a UV safe way is an open problem. However, one can just use canonical quantum gravity as an effective field theory at low energy. Gravitons then become gauge bosons that mediate gravitational interactions and the corresponding gauge invariance is local Lorentz symmetry. Taking the triangle diagram of figure 3.1 from chapter 3 and replacing the SM gauge bosons with gravitons, one derives an anomalous current. Using the Fujikawa method it is straightforward to derive the gravitational anomaly [27]

$$\partial_\mu J_L^\mu = R\widetilde{R}, \tag{15.10}$$

where the current J_L^μ is the anomalous lepton current. The departure from equilibrium occurs through inflation itself, which is what makes this mechanism paradigm shifting—it turns an enemy into a friend. Suppose the inflation is an axion field. The inflation will couple to the anomalous lepton number current gravitationally via the interaction $F(\phi)RR$, where F is an odd function in ϕ. We can write the metric of the Universe with tensor fluctuations as

$$\mathrm{d}s^2 = -\,\mathrm{d}t^2 + a^2(t)\left(\mathrm{d}x^2 + h_{ij}\mathrm{d}x^i\mathrm{d}x^j\right). \tag{15.11}$$

Gauge invariance gives us the freedom such that only two transverse traceless elements of h_{ij} are non-zero, which we denote h_L and h_R,

$$\begin{aligned}\mathrm{d}s^2 = -\,\mathrm{d}t^2 + a^2(t)\Bigg(&\left[1 - \frac{1}{\sqrt{2}}(h_L + h_R)\right]\mathrm{d}x^2\Bigg)\Bigg(+\left[1 + \frac{1}{\sqrt{2}}(h_L + h_R)\right]\mathrm{d}y^2 \\ &+ i\sqrt{2}(h_R - h_L)\mathrm{d}x\mathrm{d}y + \mathrm{d}z^2\Bigg).\end{aligned} \tag{15.12}$$

The anomalous lepton current can then be written explicitly by calculating the Ricci tensor from the above metric and contracting it with its dual

$$R\widetilde{R} = \frac{4\mathrm{i}}{a^3}\left[\partial_z^2 h_r \partial_z \partial_t h_L + a^2 \partial_t^2 h_R \partial_t \partial_z h_L + \frac{1}{2}\partial_t a^2 \partial_t h_R \partial_t \partial_L h_L - (L \leftrightarrow R)\right]. \tag{15.13}$$

From the above it is obvious that the anomalous lepton number current vanishes in the case of left–right symmetry. We need the gravitational waves to have some birefringence. This is achieved by adding the term $F(\phi)F\widetilde{F}$ to the Einstein–Hilbert action. Choosing a simple form for [28]

$$F = \frac{N\phi}{16\pi^2 M_{\mathrm{pl}}^2}, \tag{15.14}$$

where $m_{\rm pl}$ is the Planck mass and N is a dimensionless number motivated by string theory, the classical equations of motion give

$$\begin{aligned}\Box h_L &= -\,2\mathrm{i}\frac{\dot{\Theta}}{a}h_L' \\ \Box h_R &= 2\mathrm{i}\frac{\dot{\Theta}}{a}h_R',\end{aligned} \tag{15.15}$$

with

$$\Theta = \frac{N\dot{\phi}H}{2\pi^2 M_{\rm pl}^3}, \tag{15.16}$$

which indeed has a birefringence while inflation is happening. In fact one can rewrite the above equation in terms of the slow roll parameter $\epsilon = \frac{1}{2}\dot{\phi}^2/(HM_{\rm pl})^2$

$$\Theta = \sqrt{2\epsilon}\,N\frac{H^2}{2\pi^2 M_{\rm pl}^2}. \tag{15.17}$$

This elegant and relatively new idea is unfortunately well outside experimental reach, unless there is a dramatic improvement in the precision of searches for non-Gaussianities in the CMB radiation which gives us sufficiently detailed information about the inflation.

References

[1] Balázs C *et al* 2014 Baryogenesis, dark matter and inflation in the next-to-minimal supersymmetric standard model *J. High. Energy Phys.* **2014** 73

[2] Alanne T *et al* 2016 Baryogenesis in the two doublet and inert singlet extension of the Standard Model *J. Cosmol. Astropart. Phys.* JCAP **08** 057

[3] Chiang C-W, Kaori F and Eibun S 2016 Electroweak baryogenesis with lepton flavor violation *Phys. Lett.* B **762** 315–20

[4] Phong V Q, Thao N C and Long H N 2015 Baryogenesis in the Zee–Babu model *Phys. Rev.* D **97** 115008

[5] Rindler-Daller T, Li B, Shapiro P R, Lewicki M and Wells J D 2015 How scalar-field dark matter may conspire to facilitate baryogenesis at the electroweak scale *Mtg of the APS Division of Particles and Fields (DPF 2015) (Ann Arbor, MI, August 4–8)* arXiv:hep-ph-1510.08369

[6] Arcadi G, Covi L and Nardecchia M 2015 Gravitino dark matter and low-scale baryogenesis *Phys. Rev.* D **92** 115006

[7] Chao W 2015 Electroweak baryogenesis in the exceptional supersymmetric Standard Model *J. Cosmol. Astropart. Phys.* **08** 055

[8] Dorsch G C, Huber S J, Mimasu K and No J M 2014 Echoes of the electroweak phase transition: discovering a second Higgs doublet through $A_0 \to ZH_0$ *Phys. Rev. Lett.* **113** 211802

[9] Chang X and Huo R 2014 Electroweak baryogenesis in the MSSM with vectorlike superfields *Phys. Rev.* D **89** 036005

[10] Lewicki M, Rindler-Daller T and Wells J D 2016 Enabling electroweak baryogenesis through dark matter *J. High. Energy Phys.* **06** 055
[11] Dhuria M, Hati C and Sarkar U 2016 Moduli induced cogenesis of baryon asymmetry and dark matter *Phys. Lett.* B **756** 376
[12] Fuyuto K, Hisano J and Senaha E 2016 Toward verification of electroweak baryogenesis by electric dipole moments *Phys. Lett.* B **755** 491
[13] Ipek S and March-Russell J 2016 Baryogenesis via particle-antiparticle oscillations *Phys. Rev.* D **93** 123528
[14] Fukushima M, Mizuno S and Maeda K-i 2016 Gravitational baryogenesis after anisotropic inflation *Phys. Rev.* D **93** 103513
[15] Fujita T and Kamada K 2016 Large-scale magnetic fields can explain the baryon asymmetry of the Universe *Phys. Rev.* D **93** 083520
[16] Yamada M 2016 Affleck–Dine baryogenesis just after inflation *Phys. Rev.* D **93** 083516
[17] Takahashi F and Yamada M 2016 Spontaneous baryogenesis from asymmetric inflation *Phys. Lett.* B **756** 216
[18] Dev P S B and Mohapatra R N 2015 TeV scale model for baryon and lepton number violation and resonant baryogenesis *Phys. Rev.* D **92** 016007
[19] Araki T *et al* 2005 Measurement of neutrino oscillation with KamLAND: evidence of spectral distortion *Phys. Rev. Lett.* **94** 8
[20] Yanagida T 1980 Horizontal symmetry and masses of neutrinos *Prog. Theor. Phys.* **64** 3
[21] Hamaguchi K 2002 Cosmological baryon asymmetry and neutrinos: baryogenesis via leptogenesis in supersymmetric theories arXiv:hep-ph-0212305
[22] Fukugita M and Yanagida T 1986 Baryogenesis without grand unification *Phys. Lett.* B **174** 1
[23] Davidson S, Nardi E and Nir Y 2008 Leptogenesis *Phys. Rep.* **466** 4
[24] Dror J, Hiramatsu T, Kohri K, Murayama H and White G 2020 Testing the seesaw mechanism and leptogenesis with gravitational waves *Phys. Rev. Lett.* **124** 4
[25] Affleck I and Dine M 1985 A new mechanism for baryogenesis *Nucl. Phys.* B **249** 2
[26] Allahverdi R and Mazumdar A 2012 A mini review on Affleck–Dine baryogenesis *New J. Phys.* **14** 12
[27] Alvarez-Gaume L and Witten E 1984 Gravitational anomalies *Nucl. Phys.* B **234** 2
[28] Alexander S H S, Peskin M E and Sheikh-Jabbari M M 2006 Leptogenesis from gravity waves in models of inflation *Phys. Rev. Lett.* **96** 8
[29] Enqvist K and Anupam M 2003 Cosmological consequences of MSSM flat directions *Phys Rep.* **380** 99–234
[30] White G, Pearce L, Vagie D and Kusenko A 2021 Detectable gravitational wave signatures from Affleck–Dine baryogenesis *Phys. Rev. Lett.* **127** 18

Chapter 16

Discussion and outlook

Electroweak baryogenesis (EWGB) is a rich framework which attempts to partially answer one of the most fundamental questions of our existence—why is there something rather than nothing? The answer it suggests is elegant and beautiful, but its most attractive features are that it requires new physics and it is testable. The next generation of colliders and electric dipole moment (EDM) experiments should either confirm the necessary ingredients of EDMs or rule out the minimal scenario of the baryon asymmetry being produced during the transition to the Standard Model phase. Going beyond the minimal model via a multi-step electroweak phase transition [1–3] makes it possible to expand the parameter space where one expects new physics if electroweak baryogenesis is a correct history of our Universe. However, even in the multi-step scenario it remains to be seen if one can have the scale of new physics higher than what would be seen at the proposed 100 TeV collider without the scenario being contrived.

Even if nature decided that the matter–anti-matter asymmetry was indeed produced via the electroweak mechanism in a multi-step phase transition that puts the new physics scale out of our reach, there is another avenue through which EWBG could be tested. The electroweak mechanism requires that there be a strongly first-order phase transition in our cosmic history and such a phase transition leaves a relic gravitational wave background. If a phase transition occurred at either $T \in [10^7, 10^8]$ GeV, then the produced gravitational waves could be within the sensitivity ranges of LIGO, and further in the future LISA could verify a strongly first-order electroweak phase transition.

One theoretical issue we have glossed over in this analysis is the approximation that the mass basis and the vacuum expectation value vary slowly across a phase transition. The effect of the space–time varying mass matrix can be modeled through adding space–time-dependent transformation matrices, $U(x)$ to the Lagrangian [4, 5]. When we derive the equation of motion operators they now have a dependence on these transformation matrices. We proceed in the same way as before, acting on the

doi:10.1088/978-0-7503-3571-3ch16

Schwinger–Dyson equations from the left and right with the modified (that is transformation-matrix-dependent) equation of motion operators from the left and right. The sum of these two equations gives a constraint equation (which to lowest order in a suitable expansion will just give a space–time-dependent version of the usual dispersion relations) and the difference gives a kinetic equation. However, rather than taking the limit of $z \rightarrow x$, one rewrites the resulting constraint and kinetic equations in terms of the relative and center of mass coordinates—$x \mp z$, respectively—and performs the Wigner transform on the equations.

Solving the resulting integro-differential equations is a formidable task, but generic features can be teased out, in particular the important question as to whether there indeed is a resonance. Recent work involving toy models using the above-described Wigner functional approach seems to indicate that there is resonance, with the caveat that the center of the resonance may be damped. Current results seem to suggest that there is enough of a resonance behavior left over to make many models of EWBG phenomenologically viable.

References

[1] Ovanesyan S I G and Ramsey-Musolf M J 2016 Two-step electroweak baryogenesis *Phys. Rev.* D **93** 015013

[2] Patel H H, Ramsey-Musolf M J and Wise M B 2013 Color breaking in the early Universe *Phys. Rev.* D **88** 1

[3] Ramsey-Musolf M J, White G A and Winslow P 2016 Colour breaking baryogenesis *Phys. Rev.* D **97** 123509

[4] Cirigliano V *et al* 2010 Flavored quantum Boltzmann equations *Phys. Rev. D* **81** 10

[5] Cirigliano V, Lee C and Tulin S 2011 Resonant flavor oscillations in electroweak baryogenesis *Phys. Rev.* D **84** 5

www.ingramcontent.com/pod-product-compliance
Ingram Content Group UK Ltd.
Pitfield, Milton Keynes, MK11 3LW, UK
UKHW051851100926
8556IPUK00005B/577

9 780750 335690